An Open Guide to Data Structures and Algorithms

An Open Guide to Data Structures and Algorithms

PAUL W. BIBLE AND LUCAS MOSER

PALNI PRESS
INDIANAPOLIS, INDIANA

Contents

Publisher's Note

This textbook was peer-reviewed, copyedited, and published through the Private Academic Library Network of Indiana (PALNI) PALSave Textbook Creation Grants Program, which is funded by the Lilly Endowment Inc. For more information about the PALSave: PALNI Affordable Learning Program, visit the PALSave website.

If you have comments, suggestions, or corrections for this textbook, please send them to palsave@palni.edu.

PALSave

Lilly Endowment Inc.
A Private Philanthropic Foundation

Acknowledgements

We would like to thank Edward Mandity for serving as the project manager for the textbook creation process. We would also like to thank Amanda Hurford with the PALNI team for working with us when we encountered challenges along the way. We owe a great debt to our reviewers, Dr. Joshua Kiers and Dr. Aaron Boudreaux, for their helpful suggestions and key insights that drastically improved our initial draft. We would like to also thank Matthew Furber for offering graphic design advice as well as introducing us to our fabulous illustrator, Mia M. Scarlato. Additionally, we would like to thank our department chair, Dr. Matt DeLong, for his constant support of our efforts to drive learning and equity through the creation of this open-access textbook. We also wish to thank all members of Marian University's Department of Mathematical and Computational Sciences in the School of Science and Mathematics for the awesome discussions of research, pedagogy, and inclusive instruction, not to mention the fellowship and friendship.

Paul W. Bible would like to thank his wonderful wife, Dr. Colleen Doçi. Her constant love and support made this happen.

Lucas Moser would like to thank his wife and kids for supporting him in his endeavors as an educator.

1. Algorithms, Big-O, and Complexity

Learning Objectives

After reading this chapter you will...

- be able to define algorithms and data structures.
- be able to describe how this study will differ from prior academic studies.
- be able to describe how the size of the input to a procedure impacts resource utilization.
- use asymptotic notation to describe the scalability of an algorithm.

Introduction

As a student of computer science, you have already accomplished a broad array of programming tasks. In order to fully understand where we are going, let us first consider where we have been. We will start by analyzing a programming exercise from an introductory class or textbook. Consider a function that counts the occurrences of a particular character **x** within a string **s**.

```
function countOccurrences(x, s)
    set count to 0
    for each character c in s
        if c equals x
            set count to count + 1
    return count
```

With regard to learning a programming language, this exercise serves many purposes. In order to solve this problem, you must first understand various features of a language's syntax:

- Variables are units of data that allow us to store intermediate results.
- Iteration is a means of systematically visiting each element in a sequence.
- Conditionals allow us to choose whether to execute a particular set of statements.
- Functions allow us to encapsulate logic and reuse it elsewhere.

Individually, each of these concepts is neither interesting nor impactful. The real power in computation comes from combining these into meaningful solutions, and in that combination lies the purpose of this text. Moving forward, we will refer to this combination of elements as synthesis.

Synthesis is something we learn to do through the study of data structures and algorithms. These are known and well-researched solutions. This text will require us to learn the patterns and trade-offs inherent in those existing solutions. However, there is a blunt truth underlying this course of study: you will likely never implement these algorithms or data structures again. This leads to the obvious question of "Why study them in the first place?" The answer lies in the imminent transition from skills to synthesis.

Synthesis is hard, regardless of your field of study. Arithmetic, algebra, and geometry are relatively easy compared to engineering. Performing at a piano recital is not the same as composing the music. Basic spelling and grammar are not sufficient for writing a novel. Writing a conditional or loop is inherently different than creating software. There is one major difference: arithmetic, piano, and grammar all have utility outside of synthesis. There is no utility in this world for conditional statements and loops if you cannot apply them to solve new problems.

For many students, this transition from programming skills to synthesis of solutions will require significant effort. It is likely that you have experienced nothing like it before. For that reason, we choose to "stand on the shoulders of giants" to see how others have transformed simple components into elegant solutions. This will likely require us to memorize much of what we see. You may even, from time to time, be asked to recall specific details of a data structure or algorithm. However, if your study of these solutions ends there, you have certainly missed the mark.

Consider another analogy. Bootstrapping is a computing technique employed when a computer loads a program. Rather than loading all instructions needed to execute a task, the computer loads only a small number of critical instructions, and those in turn load other instructions. In a sense, the program starts out with only a seed containing the most crucial, fundamental, and useful pieces of information. That seed does not directly solve any problem but rather solves a problem by acquiring other instructions that can. The study of data structures and algorithms will bootstrap your problem-solving skills. You may or may not explicitly use anything you learned, but the ideas you have been exposed to will give you a starting point for solving new and interesting problems later.

What Is a Data Structure?

Data encountered in a computer program is classified by type. Common types include integers, floating point numbers, Boolean values, and characters. Data structures are a means of aggregating many of these scalar values into a larger collection of values.

Consider a couple of data structures that you may have encountered before. The index at the end of a book is not simply text. If you recognize it as structured data, it can help you find topics within a book. Specifically, it is a list of entries sorted alphabetically. Each entry consists of a relevant term followed by a comma-separated list of page numbers. Likewise, a deck of playing cards can be viewed as another data structure. It consists of precisely 13 values for 4 different suits, resulting in 52 cards. Each value has a specific meaning. If you wish to gain access to a single card, you may cut the deck and take whatever card is on top.

With each data structure encountered, we consider a set of behaviors or actions we want to perform on that structure. For example, we can flip over the card on the top of a deck. If we combine this with the ability to sequentially flip cards, we can define the ability to find a specific card given a deck of unsorted playing cards. Describing and analyzing processes like these is the study of algorithms, which are introduced in the next section and are the primary focus of this text.

What Is an Algorithm?

An **algorithm** is an explicit sequence of instructions, performed on data, to accomplish a desired objective. Returning to the playing card example, your professor may ask you to hand him the ace of spades from an unsorted deck. Let us consider two different procedures for how to accomplish this task.

A	B
1. Spread the cards out on a table. 2. Hand him the ace of spades.	1. Orient the deck so that it is facing 2. Inspect the value of the top card. 3. Is the card the ace of spades? a. If yes, then hand him the card. b. If no, then discard the card and r

The above is a meaningful illustration and represents our first encounter with the challenge of synthesis introduced in the first section of this chapter. When someone new to computer science is asked to find a card from a deck, the procedure more often resembles A rather than B. Whether working with cards in a deck or notes from class, our human brains can have the ability to process a reasonably sized set of unstructured data. When looking for a card in a deck or a topic in pages of handwritten notes, informal procedures like A work fine. However, they do have limitations. Procedure A lacks the precision necessary to guide a computer in solving the problem.

Algorithms and Implementations

From a design perspective, it makes sense to formulate these instructions as reusable procedures or functions. The process of reimagining an algorithm as a function or procedure is typically referred to as implementation. Once implemented, a single well-named function call to invoke a specific algorithm provides a powerful means of abstraction. Once the algorithm is written correctly, future users of the function may use the implementation to solve problems without understanding the details of the procedure. In other words, using a single high-level procedure or function to run an algorithm is a good design practice.

Expression in a Programming Language

If you are reading this book, you have at least some experience with programming. What if you are asked to express these procedures using a computer program? Converting procedure A is daunting from the start. What does it mean to spread the deck out on the table? With all "cards" strewn across a "table," we have no programmatic way of looking at each card. How do you know when you have found the ace of spades? In most programming languages there is no comparison that allows you to determine if a list of items is equal to a single value. However, procedure B can (with relative ease) be expressed in any programming language. With a couple of common programming constructs (lists, iteration, and conditionals), someone with even modest experience can typically express this procedure in a programming language.

The issue is further complicated if we attempt to be too explicit, providing details that obscure the desired intent. What if instead of saying "Then hand him the card," you say, "Using your index finger on the top of the card and your thumb on the bottom, apply pressure to the card long enough to lift the card 15 centimeters from the table and 40 centimeters to the right to hand it to your professor." In this case, your instructions have become more explicit, but at the cost of obfuscating your true objective. Striking a balance between explicit and sufficient requires practice. As a rule, pay attention to how others specify algorithms and generally lean toward being more explicit than less.

Analysis of Algorithms

Constraints

How much work does it require to find the ace of spades in a single deck of cards? How many hands do you need? An algorithm (such as procedure B) will always be executed in the presence of constraints. If you are searching for the ace of spades, you would probably like your search to terminate in a reasonable amount of time. Thirty seconds to find the card is probably permissible, but four hours likely is not. In addition, finding the card in the deck requires manipulation of the physical cards with your hands. If you had to perform this task with one hand in your pocket, you likely would have to consider a different algorithm.

Constraints apply in the world of software as well. Even though modern CPUs can perform operations at an incredible rate, there is still a very real limit to the operations that can be performed in a fixed time span. Computers also have a fixed amount of memory and are therefore space constrained.

Scalability

A primary focus in the study of algorithms is what happens when we no longer have a single deck of 52 cards. What if we have a thousand decks and really bad luck (all 1,000 aces of spades are at the back of the deck)? No human hands can simultaneously hold that many cards, and even if they could, no human is going to look at 51,000 other cards to check for an ace of spades. This illustrates the desire for algorithms to be scalable. If we start with a problem of a certain size (say, a deck of 52 cards), we particularly care about *how much harder it is to perform the procedure with two decks.* With a large enough input, any algorithm will eventually become impractical. In other words, we will eventually reach either a space constraint or a time constraint. It is therefore necessary that realistic problem sizes be smaller than those thresholds.

Measuring Scalability

Measuring scalability can be challenging. It seems obvious (but worth noting) that a given algorithm for sorting integers may be able to sort 10 trillion values on a distributed supercomputer but will likely not terminate on a mobile phone in a reasonable amount of time. This means the simple model for measuring scalability (namely, how long an algorithm takes to run) is insufficient. Comparing algorithms without actually running them on a physical machine would be useful. This is the main topic of this section.

If we wish to analyze algorithm performance without actually executing it on a computer, we must define some sort of abstraction of a modern computer. There are a number of models we could choose to work with, but the most relevant for this text is the uniform cost model. With this model, we choose to assume that any operation performed takes a uniform amount of time. In many real-world scenarios, this may likely be inaccurate. For example, in every coding framework and every machine architecture, multiplication is always faster than division. So why is it acceptable to assume a uniform cost? When learning algorithms and data structures for the first time, assuming uniform cost usually gives us a sufficiently granular impression of runtime while also retaining ease of understanding. This allows us to focus on the synthesis and implementation of algorithms and deal with more precise models later.

One last aspect of modeling is that we need not measure all operations in all cases. It is quite common, for example, to only consider comparison operations in sorting algorithms. This is justifiable in many cases. Consider comparing one algorithm or one implementation to another. We may be able to employ some clever tricks to marginally improve performance. However, for most general-purpose sorting algorithms, the number of comparisons is the most meaningful operation to count. In some cases, it may make sense to count the number of times an object was referenced or how many times we performed addition. Keep in mind, the important part of algorithm analysis is how many more operations we have to do each time we increase the size of our input. Fortunately, we have a notation that helps us describe this growth. In the next section, we will formally define asymptotic notation and observe how it is helpful in describing the performance of algorithms.

Algorithm Analysis

When comparing algorithms, it is typically not sufficient to describe how one algorithm performs for a given set of inputs. We typically want to quantify *how much better one algorithm performs when compared to another for a given set of inputs.*

To describe the cost of a software function (in terms of either time or space), we must first represent that cost using a mathematical function. Let us reconsider finding the ace of spades. How many cards will we have to inspect? If it is the top card in the deck, we only have to inspect 1. If it is the bottom card, we will have to inspect 52. Immediately, we start to see that we must be more precise when defining this function. When analyzing algorithms, there are three primary cases of concern:

- Worst case: This function describes the most comparisons we may have to make given the current algorithm. In our ace of spades example, this is represented by the scenario when the ace of spades was the bottom card in the deck. If we wish to represent this as a function, we may define it as $f(n) = n$. In other words, if you have 52 cards, the most comparisons you will have to perform will be 52. If you add two jokers and now have 54 cards, you now have to perform at most 54 comparisons.

- Average case: This function describes what we would typically have to do when performing an algorithm many times. For example, imagine your professor asked you to first find the ace of spades, then 2 of spades, then 3 of spades, until you have performed the search for every card in the deck. If you left the card in the deck each round, some cards would be near the top of the deck and others would be near the end. Eventually, each iteration would have a cost function of roughly $f(n) = n/2$.
- Amortized case: This is a challenging case to explain early in this book. Essentially, it arises whenever you have an expensive set of operations that only occur sometimes. We will encounter this in the resizing of hash tables as well as the prerequisites for Binary Search.

You may wonder why best-case scenarios are not considered. In most algorithms, the best case is typically a small, fixed cost and is therefore not very useful when comparing algorithms.

Consider a very literal interpretation of a uniform cost model for our ace of spades example. We would then have three different operations:

- compare (C): comparing a card against the desired value
- discard (D): discarding a card that does not match
- return (R): handing your professor the card that does match

Under this model, our cost function will be the following under the worst-case scenario, where n is the number of cards in the deck:

$$f(n) = nC + nD + R.$$

Because we consider all costs to be uniform, we can set C, D, and R all to some constant. To make our calculations easier, we will choose 1. Our cost function is now slightly easier to read:

$$f(n) = 1n + 1n + 1$$

$$f(n) = 2n + 1.$$

This implies that, regardless of how many cards we have in our deck, we have a small cost of 1 (namely, the amount of time to return the matching card). However, if we add two jokers, we have increased our deck size by 2, thus increasing our cost by 4. Notice that the first element in our function is variable (2n changes as n changes), but the second is fixed. As our n gets larger, the coefficient of 2 on 2n has a much greater impact on the overall cost than the cost to return the matching card.

Big-O Notation

Imagine now that another student in your class developed a separate algorithm, which has a cost function of $m(n) = n^2 + 1$. The question is now, Which algorithm performs better than the other as we add more cards to the deck? Consider the graph of f and m below, where a shallower slope represents a slower growth rate.

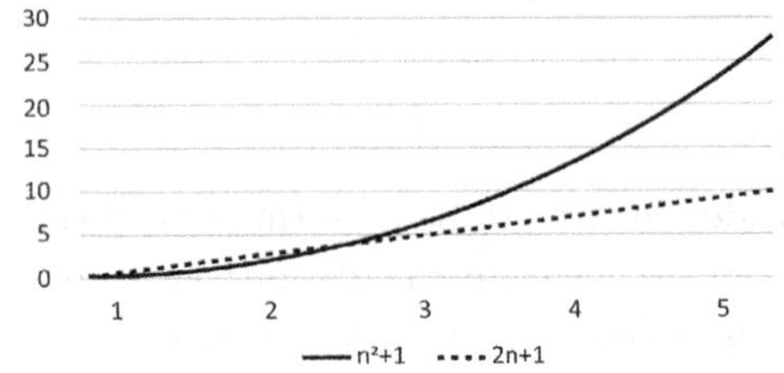

Figure 1.1

Note that when n = 1, your classmate's algorithm performs fewer operations than our original. However, for all n values that come after 2, the cost of our original algorithm wins outright. Although this does give us some intuition about the performance of algorithms, it would behoove us to define more precise notation before moving forward.

Asymptotic notation gives us a way of describing how the output of a function grows as the inputs become bigger. We will address three different notations as part of this chapter, but the most important is Big-O. Most often stylized with a capital O, this notation enables us to classify cost functions into various well-known sets. Formally, we define Big-O as follows:

$f(n) = O(g(n))$ if $f(n) \le cg(n)$ for some $c > 0$ and all $n > n_0$.

While this is a very precise and useful definition, it does warrant some additional explanation:

- O(g(n)) is a set of all functions that satisfy the condition that they are "less than" some constant multiplied by g(n). While it is conventional to say that f(n) = O(g(n)), it is more accurate to read this as "f(n) is a member of O(g(n))."
- When determining whether a function is a member of O(g(n)), any positive real number may be chosen for c.
- The most important aspect of this notation is that the inequality holds when n is really big. As a result, asserting that it holds for relatively small n values is not necessary. We can choose an n_0. Then for all values greater than it, our inequality must hold.

Through some basic algebra, we can determine that f(n) is O(n), and m(n) is $O(n^2)$:

f(n)=2n + 1 and g(n) = n, then we can show f(n) = O(n) as follows:

$$2n + 1 \leq cn$$

$$2n + 1 \leq 3n$$

$$1 \leq n$$

$n \geq 1$the original definition holds when $n_0 = 1$.

$m(n)=n^2+1$ and $g(n) = n^2$, then we can show $m(n) = O(n^2)$

$$n^2 + 1 \leq cn^2$$

$$n^2 + 1 \leq 2n^2$$

$$1 \le n^2$$

$$n^2 \ge 1$$

$n \ge 1$the original definition holds when $n_0 = 1$.

Notably, it is possible to show that $f(n) = O(n^2)$. However, it is not true that $m(n) = O(n)$. Showing this has been left as an exercise at the end of the chapter.

Another Example

Someone new to algorithm analysis may start to draw conclusions that cost functions with higher degree polynomials are unequivocally slower (or worse) than those with lower degree polynomials. Consider the following two cost functions and what they look like when n is less than 10:

$$s(n) = 64n$$

$$t(n) = n^2.$$

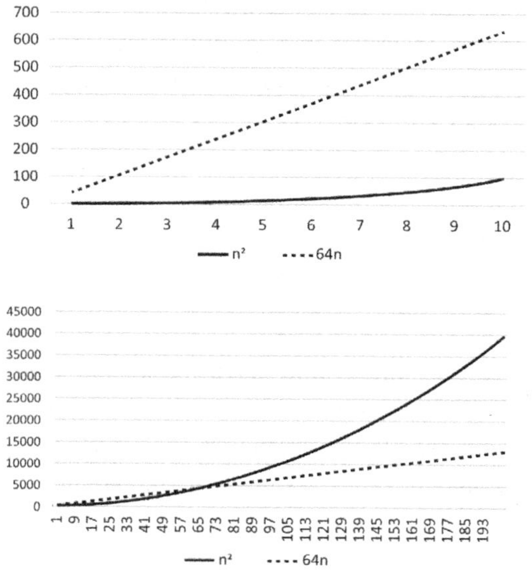

Figure 1.2

As the size of your input (n in this case) gets larger, higher degree terms will always have more influence over the growth of the function than large coefficients on smaller degree terms. It is important though that we do not immediately conclude that the algorithm for function s is inherently better than or more useful than the algorithm for function t. Notice that for small values of n (specifically, those smaller than 64), algorithm t actually outperforms s. There are real-world scenarios where input size is known to be small, and an asymptotically less-than-ideal algorithm may actually be preferred due to other desirable attributes.

Big-O notation captures the asymptotic scaling behavior of an algorithm. This means the resource costs grow as n goes to infinity. It is seen as a measure of the "complexity" of an algorithm.

In this text, we may refer to the "time complexity" of an algorithm, which means its Big-O worst-case scaling behavior. If one algorithm runs in O(n) time and the other in $O(n^2)$ time, we may say that the O(n) algorithm is an order of magnitude faster than the $O(n^2)$ algorithm. The same terms may also be applied to assessments of the space usage of an algorithm.

Other Notations

The remainder of this book (and most books for that matter) typically uses Big-O notation. However, other sources often reference Big-Theta notation and a few also use Big-Omega. While we will not use either of these extensively, you should be familiar with them. Additional notations outside these three do exist but are encountered so infrequently that they need not be addressed here.

Big-Omega Notation

Whereas Big-O notation describes the upper bound on the growth of a function, Big-Omega notation describes the lower bound on growth. If you describe Big-O notation as "some function f will never grow faster than some other function g," then you could describe Big-Omega as "some function f will never grow more slowly than g." Herein lies the explanation of why Big-Omega does not see much usage in real-world scenarios. In computer science, upper bounds are typically more useful than lower bounds when considering how an algorithm will perform on a large scale. In other words, if an algorithm performs better than expected, we are pleasantly surprised. If it performs worse, it may take a long time, if ever, to complete. The formal definition of Big-Omega is as follows:

$$f(n) = \Omega(g(n)) \text{ if } c * g(n) \le f(n) \text{ for some } c > 0 \text{ and all } n > n_0.$$

Big-Theta Notation

Big-Theta of a function, stylized as $\theta(g(n))$, has more utility in routine analysis than Big-Omega. Recall that $f(n) = 2n+1$ is both $O(n)$ and $O(n^2)$. This is true because the growth of the function has an upper bound of n as well as n^2. Although it is possible to reference different functions, it is common to choose the slowest-growing g(n) such that g(n) is an upper bound on function f. We can see that Big-O may not be as precise as we would like, and this can result in some confusion. Therefore, in this book, we will be interested in the smallest g(n) that can serve to bound the algorithm at O(g(n)). As a result of the ambiguity of Big-O, some computer scientists prefer to use Big-Theta. In this notation, we choose to specify both the upper and lower bounds on growth by using a single function g(n). This removes confusion and allows for a more precise description of the growth of a function. The definition of Big-Theta is as follows:

$$f(n) = \theta(g(n)) \text{ if } c_1 * g(n) \le f(n) \le c_2 * g(n) \text{ for some } c_1 > 0, c_2 > 0 \text{ and } n > n_0.$$

It should be noted that not all algorithms can be described

using Big-Theta notation, as it requires that an algorithm be bounded from above and below by the same function.

Exercises

1. Consider a new task for a deck of cards. Someone has cut the deck (split it into two parts), flipped one part upside down, then shuffled the deck back together. As a result, we currently have one deck that has some cards facing up, others facing down, and no discernable pattern to predict which is up or down. Write a description (algorithm) for how to put all cards face up in the deck. Be precise enough for your procedure to be reproducible but not so verbose that the reader loses track of the core components of the algorithm.
2. Write a cost function (using a uniform cost model) that describes the work necessary to reorient a deck of n cards. If your deck is suddenly m cards larger, how much additional work must be completed?
3. For each function below, specify whether it is $O(n)$, $O(n^2)$, or both.

   a. $f(n) = 8n + 4n$
   b. $f(n) = n(n+1)/2$
   c. $f(n) = 12$
   d. $f(n) = 1000n^2$

References

Cormen, Thomas H., Charles E. Leiserson, Ronald L. Rivest, and Clifford Stein. *Introduction to Algorithms*, 2nd ed. Cambridge, MA: The MIT Press, 2001.

2. Recursion

Learning Objectives

After reading this chapter you will...

- understand the features of recursion and recursive processes.
- be able to identify the components of a recursive algorithm.
- be able to implement some well-known recursive algorithms.
- be able to estimate the complexity of recursive processes.
- understand the benefits of recursion as a problem-solving strategy.

Introduction

Recursion is a powerful tool for computation that often saves the programmer considerable work. As you will see, the benefit of recursion lies in its ability to simplify the code of algorithms, but first we want to better understand recursion. Let's look at a simple nonrecursive function to calculate the product of 2 times a nonnegative integer, n, by repeated addition:

```
function multiplyBy2(n)
  set sum to 0
  for i = 1 to n
    set sum to sum + 2
  return sum
```

This function takes a number, n, as an input parameter and then defines a procedure to repeatedly add 2 to a sum. This approach to calculation uses a for-loop explicitly. Such an approach is sometimes referred to as an iterative process. This means that the calculation repeatedly modifies a fixed number of variables that change the current "state" of the calculation. Each pass of the loop updates these state variables, and they evolve toward the correct value. Imagine how this process will evolve as a computer executes the function call **multiplyBy2(3)**. A "call" asks the computer to evaluate the function by executing its code.

When the process starts, **sum** is 0. Then the process iteratively adds 2 to the **sum**. The process generated is equivalent to the mathematical expression (2 + 2 + 2). The following table shows the value of each variable (**i**, **sum**, and **n**) at each time step:

Time	variable **i**	variable **sum**	variable **n**
Time 0 (before the loop)	Not initialized	0	3
Time 1	1	2	3
Time 2	2	4	3
Time 3	3	6	3

Figure 2.1

Recursive Multiplication

Like iterative procedures, recursive procedures are a means to repeat certain operations in code. We will now write a recursive function to calculate the multiplication by 2 as a sequence of addition operations.

```
function recursiveMultiplyBy2(n)
    if n equals 0
        return 0
    else
        return 2 + recursiveMultiplyBy2(n - 1)
```

The recursive formulation follows the mathematical intuition that 2 * n = 2 + 2 * (n - 1) = 2 + 2 + 2 * (n - 2) ... and so on until you reach 2 + 2 + 2 + ... 2 * 1. We can visualize this process by considering how a computer might evaluate the function call **recursiveMultiplyBy2(3)**. The evaluation process is similar conceptually to a rewriting process.

recursiveMultiplyBy2(3) -> 2 + recursiveMultiplyBy2(2)

-> 2 + 2 + recursiveMultiplyBy2(1)

-> 2 + 2 + 2 + recursiveMultiplyBy2(0)

-> 2 + 2 + 2 + 0

-> 2 + 2 + 2

-> 6

This example demonstrates a few features of a recursive procedure. Perhaps the most recognizable feature is that it makes a call to itself. We can see that **recursiveMultiplyBy2** makes a call to **recursiveMultiplyBy2** in the else part of the procedure. As an informal definition, a recursive procedure is one that calls to itself (either directly or indirectly). Another feature of this recursive procedure is that the action of the process is broken into two parts. The first part directs the procedure to return 0 when the input, n, is equal to 0. The second part addresses the other case where n is not

0. We now have some understanding of the features of all recursive procedures.

Features of Recursive Procedures

- A recursive procedure makes reference to itself as a procedure. This self-call is known as the recursive call.
- Recursive procedures divide work into two cases based on the value of their inputs.
  - One case is known as the base case.
  - The other case is known as the recursive case or sometimes the general case.

Recursive Exponentiation

Let us now consider another example. Just as multiplication can be modeled as repeated addition, exponentiation can be modeled as repeated multiplication. Suppose we wanted to modify our iterative procedure for multiplying by 2 to create an iterative procedure for calculating powers of 2 for any nonnegative integer n. How might we modify our procedure? We might simply change the operation from + (add) to * (multiply).

```
function powerOf2(n)
    set product to 1
    for i from 1 to n
        set product to product * 2
    return product
```

This procedure calculates a power of 2 for any nonnegative integer n. We changed not only the operation but also the starting value from 0 to 1.

We can also write this procedure recursively, but first let us think about how to formulate this mathematically by looking at an example. Suppose we want to calculate 2^3:

$$2^3=2 * 2^2$$
$$=2 * 2 * 2^1$$
$$=2 * 2 * 2 * 2^0$$
$$=2 * 2 * 2 * 1=8.$$

The explicit 2^0 is shown to help us think about the recursive and base cases. From this example, we can formulate two general rules about exponentiation using 2 as the base:

$$2^0=1$$
$$2^n=2 * 2^{(n-1)}.$$

Now let's write the recursive procedure for this function. Try to write this on your own before looking at the solution.

```
function recursivePowerOf2(n)
    if n equals 0
        return 1
    else
        return 2 * recursivePowerOf2(n - 1)
```

The Structure of Recursive Algorithms

As mentioned above, recursive procedures have a certain structure that relies on self-reference and splitting the input into cases based on its value. Here we will discuss the structure of recursive procedures and give some background on the motivation for recursion.

Before we begin, recall from chapter 1 that a procedure can be thought of as a specific implementation of an algorithm. While these are indeed two separate concepts, they can often be used interchangeably when considering pseudocode. This is because the use of an algorithm in practice should be made as simple as possible. Often this is accomplished by wrapping the algorithm in a simple procedure. This design simplifies the interface, allowing the programmer to easily use the algorithm in some other software context such as a program or application. In this chapter, you can treat algorithm and procedure as the same idea.

Some Background on Recursion

The concept of recursion originated in the realm of mathematics. It was found that some interesting mathematical sequences could be defined in terms of themselves, which greatly simplified their definitions. Take for example the Fibonacci sequence:

$$\{0, 1, 1, 2, 3, 5, 8, 13, 21, 34, 55, \ldots\}.$$

This sequence is likely familiar to you. The sequence starts at 0, 0 is followed by 1, and each subsequent value in the sequence is derived from the previous two elements in the sequence. This seemingly simple sequence is fairly difficult to define in an explicit way. Let's look a bit closer.

Based on the sequence above, we see the following equations are true:

$$F_0 = 0$$

$$F_1 = 1$$

$$F_2 = 1$$

$$F_3 = 2$$

$$F_4 = 3.$$

Using this formulation, we may wish to find a function that would calculate the nth Fibonacci number given any positive integer, n:

$$F_n = ?.$$

Finding such a function that depends only on n is not trivial. This also leads to some difficulties in formally defining the sequence. To partially address this issue, we can use a recursive definition. This allows the sequence to be specified using a set of simple rules that are self-referential in nature. These recursive rules are referred to in mathematics as recurrence relations. Let's look at the recurrence relation for the Fibonacci sequence:

$$F_0=0$$

$$F_1=1$$

$$F_n = F_{n-1} + F_{n-2}.$$

This gives a simple and perfectly correct definition of the sequence, and we can calculate the nth Fibonacci number for any positive integer n. Suppose we wish to calculate the eighth Fibonacci number. We can apply the definition and then repeatedly apply it until we reach the F_0 and F_1 cases that are explicitly defined:

$$F_8 = F_{8-1} + F_{8-2}$$

$$= F_7 + F_6$$

$$= (F_{7-1} + F_{7-2}) + (F_{6-1} + F_{6-2})$$

...and so on.

This may be a long process if we are doing this by hand. This could be an especially long process if we fail to notice that F_{7-2} is the same as F_{6-1}.

A Trade-Off with Recursion

Observing this process leads us to another critical insight about recursive processes. What may be simple to describe may not be efficient to calculate. This is one of the major drawbacks of recursion in computing. You may be able to easily specify a correct algorithm using recursion in your code, but that implementation may be wildly inefficient. Recursive algorithms can usually be rewritten to be more efficient. Unfortunately, the efficiency of the implementation comes with the cost of losing the simplicity of the recursive code. Sacrificing simplicity leads to a more difficult implementation. Difficult implementations allow for more bugs.

An Aside on Navigating the Efficiency-Simplicity Trade-Off

In considering the trade-off between efficiency and simplicity, context is important, and there is no "right" answer. A good guideline is to focus on correct implementations first and optimize when there is a problem (verified by empirical tests). Donald Knuth references Tony Hoare as saying "premature optimization is the root of all evil" in programming. This should not be used as an excuse to write inefficient code. It takes time to write efficient code. "I was optimizing the code" is a poor excuse for missed deadlines. Make sure the payoff justifies the effort.

Recursive Structure

The recursive definition of the Fibonacci sequence can be divided into two parts. The first two equations establish the first two values of the sequence explicitly by definition for $n = 0$ and $n = 1$. The last equation defines Fibonacci values in the general case for any integer $n > 1$. The last equation uses recursion to simplify the general case. Let's rewrite this definition and label the different parts.

Base Cases

$$F_0 = 0$$

$$F_1 = 1$$

Recursive Case

$$F_n = F_{n-1} + F_{n-2}$$

Recursive algorithms have a similar structure. A recursive procedure is designed using the base case (or base cases) and a recursive case. The base cases may be used to explicitly define the output of a calculation, or they may be used to signal the end of a recursive process and stop the repeated execution of a procedure. The recursive case makes a call to the function itself to solve a portion of the original problem. This is the general structure of a recursive algorithm. Let's use these ideas to formulate a simple algorithm to calculate the nth Fibonacci number. Before we begin, let's write the cases we need to consider.

Base Cases

n = 0 the algorithm should return 0

n = 1 the algorithm should return 1

Recursive Case

any other n the algorithm returns the sum of Fibonacci(n – 1) and Fibonacci(n – 2)

Now we define the recursive algorithm to calculate the nth Fibonacci number.

```
function fibonacci(n)
  if n equals 0
    return 0
  else if n equals 1
    return 1
  else
    return fibonacci(n − 1) + fibonacci(n − 2)
```

In this algorithm, conditional if-statements are used to select the appropriate action based on the input value of n. If the value is 0 or 1, the input is handled by a base case, and the function directly returns the appropriate value. If another integer is input, the recursive case handles the calculation, and two more calls to the procedure are executed. You may be thinking, "Wait, for every function call, it calls itself two more times? That doesn't sound efficient." You would be right. This is not an efficient way to calculate the Fibonacci numbers. We will revisit this idea in more detail later in the chapter.

Before we move on, let us revisit the two examples from the introduction, multiplication and exponentiation. We can define these concepts using recurrence relations too. We will use the generic letter a for these definitions.

Multiplication by 2

Base Case

$$a_0 = 0$$

Recursive Case

$$a_n = 2 + a_{n-1}$$

Powers of 2

Base Case

$$a_0 = 1$$

Recursive Case

$$a_n = 2 * a_{n-1}$$

Looking back at the recursive functions for these algorithms, we see they have a shared structure that uses conditionals to select the correct case based on the input, and the base cases and recursive cases handle the calculation by ending the chain of recursive calls or by adding to the chain of calls respectively. In the next section, we will examine some important details of how a computer executes these function definitions as dynamic processes to generate the correct result of a calculation.

Recursion and the Runtime Stack

Thus far, we have relied on your intuitive understanding of how a computer executes function calls. Unless you have taken a computer organization or assembly language course, this process may seem a little mysterious. How does the computer execute the same procedure and distinguish the call to **recursiveMultiplyBy2(3)** from the call to **recursiveMultiplyBy2(2)** that arises from the previous function call? Both calls utilize the same definition, but one has the value 3 bound to **n** and the other has the value 2 bound to **n**. Furthermore, the result of **recursiveMultiplyBy2(2)** is needed by the call to **recursiveMultiplyBy2(3)** before a value may be returned. This requires that a separate value of n needs to be stored in memory for each execution of the function during the evaluation process. Most modern computing systems utilize what is known as a runtime stack to solve this problem.

A stack is a simple data structure where data are placed on the top and removed from the top. Like a stack of books, to get to a book in the middle, one must first remove the books on top. We will address stacks in more detail in a later chapter, but we will briefly introduce the topic. Here we will examine how a runtime stack is used to store the necessary data for the execution of a recursive evaluation process.

The Runtime Stack

We will give a rough description of the runtime stack. This should be taken as a cartoon version or caricature of the actual runtime stack. We encourage you to supplement your understanding of this topic by exploring some resources on computer organization and assembly language. The real-world details of computer systems can be as important as the abstract view presented here.

The runtime stack is a section of memory that stores data associated with the dynamic execution of many function calls. When a function is called at runtime, all the data associated with that function call, including input arguments and locally defined variables, are placed on the stack. This is known as "pushing" to the stack. The data that was pushed onto the stack represents the state of the function call as it is executing. You can think of this as the function's local environment during execution. We call this data stored on the stack a **stack frame**. The stack frame is a contiguous section of the stack that is associated with a specific call to a procedure. There may be many separate stack frames for the same procedure. This is the case with recursion, where the same function is called many times with distinct inputs.

Once the execution of a function completes, its data are no longer needed. At the time of completion, the function's data are removed from the top of the stack or "popped." Popping data from the runtime stack frees it, in a sense, and allows that memory to be used for other function calls that may happen later in the program execution. As the final two steps in the execution of the procedure, the function's stack frame is popped, and its return value (if the function returns a value) is passed to the caller of the function. The calling function may be a "main" procedure that is driving the program, or it may be another call to the same function as in recursion. This allows the calling function to proceed with its execution. Let's trace a simple recursive algorithm to better understand this process.

A Trace of a Simple Recursive Call

Suppose we are running a program that makes a call to **recursiveMultiplyBy2(3)**. Consider the simple program below.

```
1  function recursiveMultiplyBy2(n)
2      if n equals 0
3          return 0
4      else
5          return 2 + recursiveMultiplyBy2(n-1)
6
7  function main()
8      set result to recursiveMultiplyBy2(3)
9      print result
```

The computer executes programs in a step-by-step manner, executing one instruction after another. In this example, suppose that the computer begins execution on line 8 at the start of the **main** procedure. As **main** needs space to store the resulting value, we can imagine that a place has been reserved for **main**'s result variable and that it is stored on the stack. The following figure shows the stack at the time just before the function is called on line 8.

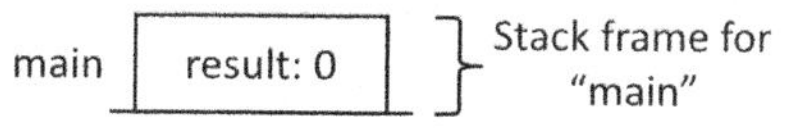

Figure 2.2

When the program reaches our call to **recursiveMultiplyBy2(3)**, the n variable for this call is bound with the 3 value, and these data are pushed onto the stack. The figure below gives the state of the stack just before the first line of our procedure begins execution.

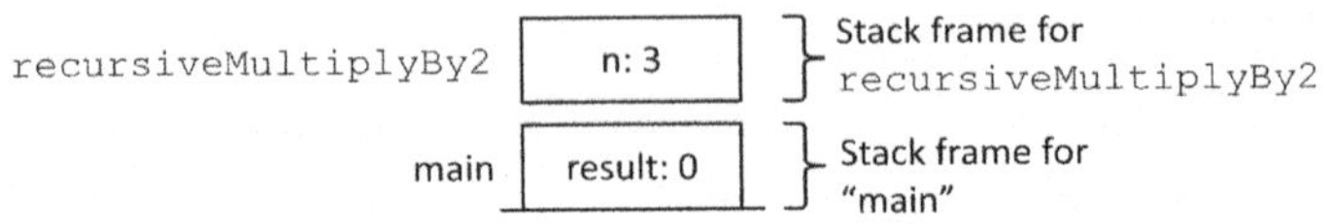

Figure 2.3

Here the value of 3 is bound to **n** and the execution continues line by line. As this **n** is not 0, the execution would continue to line 5, where another recursive call is made to **recursiveMultiplyBy2(2)**. This would push another stack frame onto the stack with 2 bound to **n**. This is shown in the following figure.

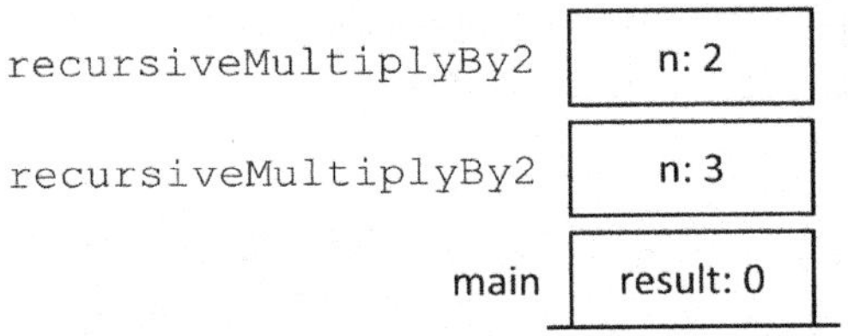

Figure 2.4

You can imagine this process continuing until we reach a call that would be handled by the base case of our recursive algorithm. The base case of the algorithm acts as a signal that ends the recursion. This state is shown in the following figure:

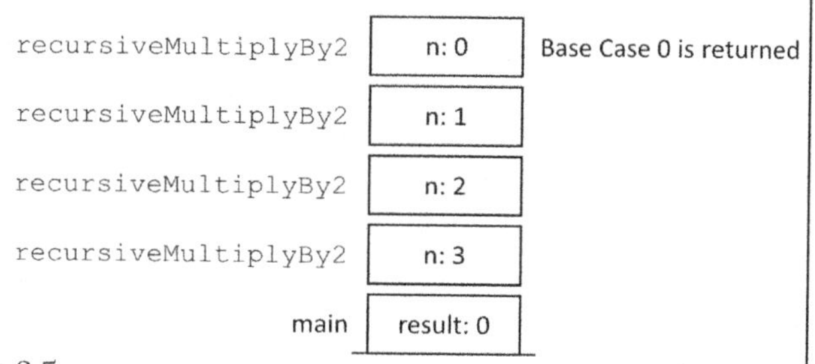

Figure 2.5

Next, the process of completing the calculation begins. The resulting value is returned, and the last stack frame is popped from the top of the stack.

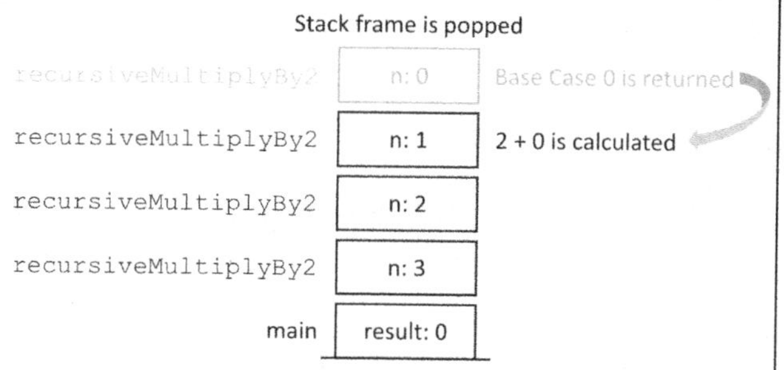

Figure 2.6

Once the call for **n = 0** gets the returned value of 0, it may then complete its execution of line 5. This triggers the next pop from the stack, and the value of 2 + 0 = 2 is returned to the prior call with n = 2. Let's look at the stack again.

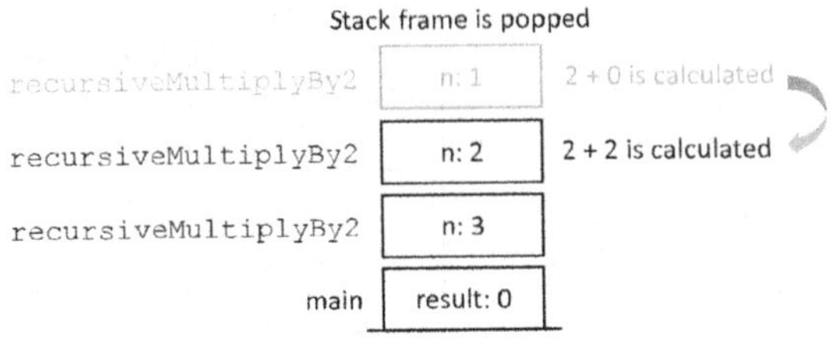

Figure 2.7

As we can see, the chain of calls is unwinding as the results are calculated in turn. This process continues. As the results are calculated and stack frames are popped, it should be clear that less memory is being used by the program. It should be noted that this implies a memory cost for deep chains of recursion. Finally, the last recursive call completes its line 5, and the value is returned to the main function as shown below.

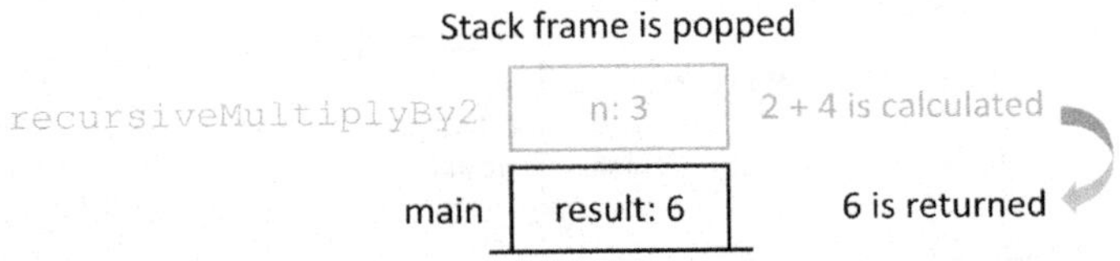

Figure 2.8

Now the main function has the result of our recursive algorithms. The value 6 is bound to the "result" variable, and when the program executes the print command, "6" would appear on the screen. This concludes our trace of the dynamic execution of our program.

Why Do We Need to Understand the Runtime Stack?

You may now be thinking, "This seems like a lot of detail. Why do we need to know all this stuff?" There are two main reasons why we want you to better understand the runtime stack. First, it should be noted that function calls are not free. There is overhead involved in making a call to a procedure. This involves copying data to the stack, which takes time. The second concern is that making function calls, especially recursive calls, consumes memory. Understanding how algorithms consume the precious resources of time and space is fundamental to understanding computer science. We need to understand how the runtime stack works to be able to effectively reason about memory usage of our recursive algorithms.

More Examples of Recursive Algorithms

We have already encountered some examples of recursive algorithms in this chapter. Now we will discuss a few more to understand their power.

Recursive Reverse

Suppose you are given the text string "HELLO," and we wish to print its letters in reverse order. We can construct a recursive algorithm that prints a specific letter in the string after calling the algorithm again on the next position. Before we dive into this algorithm, let's explain a few conventions that we will use.

First, we will treat strings as lists or arrays of characters. Characters are any symbols such as letters, numbers, or other type symbols like "*," "+," or "&." Saying they are lists or arrays means that we can access distinct positions of a string using an index. For example, let **message** be a string variable, and let its value be "HELLO." If we access the elements of this array using zero-based indexing, then **message[0]** is the first letter of the string, or "H." We will be using **zero-based indexing** (or 0-based indexing) in this textbook. Switching between 0-based and 1-based indexing should be easy, although it can be tricky and requires some thought when converting complex algorithms. Next, saying that a string is an array is incorrect in most programming languages. An array is specifically a fixed-size, contiguous block of memory designed to store multiple values of the same type. Most programming languages provide a string type that is more robust. Strings are usually represented as data structures that provide more functionality than just a block of memory. We will treat strings as a data structure that is like an array with a little more functionality. For example, we will assume that the functionality exists to determine the size of a string from its variable in constant time or O(1). Specifically, we will use the following function.

```
function length(array)
    return the length of the array or string
```

Back to the algorithm, we will define the base and recursive cases. The base case that terminates the algorithm will be reached by the algorithm when the position to print is greater than or equal to the length of the string. The recursive case will call the algorithm for the next position and then print the letter at the current position *after* the recursive call.

```
1 function recReverse(message, position)
2     if position >= length(message)
3         # return from the function, do nothing
4         return
5     else
6         # make a recursive call
7         recReverse(message, position + 1)
8         print message[position]
```

Calling **recReverse("HELLO", 0)** should give the following text printed to the screen:

O

L

L

E

H

This demonstrates that the recursive algorithm can print the characters of a string in reverse order without using excessive index manipulation. Notice the order of the print statement and the recursive call. If the order of lines 7 and 8 were switched, the characters would print in their normal order. This algorithm resembles another recursive algorithm for visiting the nodes of a tree data structure that you will see in a later chapter.

Wrappers and Helper Functions for Recursion

You may have noticed that **recReverse("HELLO", 0)** seems like an odd interface for a function that reverses a string. There is an extra piece of state, the value 0, that is needed anytime we want to reverse something. Starting the reverse process in the middle of the string at index 3, for example, seems like an uncommon use case. In general, we would expect that reversing the string will almost always be done for the entire string. To address this problem, we will make a helper function. Let's create this helper function so we can call the reverse function without giving the 0 value.

```
function reverse(message)
    recReverse(message, 0)
```

Now we may call **reverse("HELLO")**, which in turn, makes a call that is equivalent to **revReverse("HELLO", 0)**. This design method is sometimes called **wrapping**. The recursive algorithm is wrapped in a function that simplifies the interface to the recursive algorithm. This is a very common pattern for dealing with recursive algorithms that need to carry some state of the calculation as input parameters. Some programmers may call **reverse** the wrapper and **recReverse** the helper. If your language supports access specifiers like "public" and "private," you should make the wrapper a "public" function and restrict the helper function by making it "private." This practice prevents programmers from accidentally mishandling the index by restricting the use of the helper function.

Greatest Common Divisor

An algorithm for the greatest common divisor (GCD) of two integers can be formulated recursively. Suppose we have the fraction 16/28. To simplify this fraction to 4/7, we need to find the GCD of 16 and 28. The method known as Euclid's algorithm solves this problem by dividing two numbers, taking their remainder after division, and repeating the division step with one of the numbers and the remainder. Given two integers a and b, the algorithm proceeds according to this general process:

1. Take two integers a and b, and let a be larger than b.
2. Find the remainder of a / b, and let that number be r.
3. If r is 0, b is the GCD; otherwise, repeat the process with b and r in place of a and b in step 1.

The equations below illustrate this process. Here, let a equal 28, and let b equal 16:

a / b

28 / 16=1 with remainder 12

16 / 12=1 with remainder 4

12 / 4=3 with remainder 0

Since the remainder is 0, the GCD is 4.

For the fraction 16/28, dividing the denominator and the numerator by 4 gives the simplified fraction 4/7. This is illustrated below:

(16 / 4) / (28 / 4) = 4 / 7.

Before we start writing the algorithm, let us think about the base case and the recursive case. What signals the end of recursion? If the remainder of a divided by b is zero, this means that b is the greatest common divisor. This should be our base case. With any other inputs, the algorithm should make a recursive call with updated inputs. Let us examine one way to implement this algorithm. We will use the keyword **mod** to mean the remainder of two integers. For example, we will take the expression **7 mod 3** to be evaluated to the value 1. The keyword **mod** is a shortened version of the word "modulo," which is the operation that calculates integer remainders after division.

```
function GCD(a, b)
    set remainder to a mod b
    if remainder equals 0
        return b
    else
        return GCD(b, remainder)
```

This process will continue to reduce a and b in sequence until the remainder is 0, ultimately finding the greatest common divisor. This provides a good example of a recursive numerical algorithm that has a practical use, which is for simplifying fractions.

Recursive Find Minimum

Finding the minimum value in a collection of numbers can be formulated as a recursive algorithm. Let us use an array of integers for this algorithm. Our algorithm will use the helper and wrapper pattern, and it will use an extra parameter to keep track of the current minimum value. Let us begin by specifying the core algorithm as a helper function. Here, as in **recReverse**, we assume a **length** function can provide the length of the array:

```
function recMinHelper(array, position, currentMin)
    if position >= length(array)
        return currentMin
    else
        set testValue to array[position]
        if testValue < currentMin
            return recMinHelper(array, position + 1, testValue)
        else
            return recMinHelper(array, position + 1, currentMin)

function recMin(array)
    set startValue to array[0]
    return recMinHelper(array, 0, startValue)
```

For this algorithm, our base case occurs when the recursive process reaches the last position of the array. This signals the end of recursion, and the **currentMin** value is returned. In the recursive case, we compare the value of the current minimum with the value at the current position. Next, a recursive call is made that uses the appropriate current minimum and increases the position.

Recursive Algorithms and Complexity Analysis

Now that we have a better understanding of recursive algorithms, how can their complexity be evaluated? Computational complexity is usually evaluated in terms of time complexity and space complexity. How does the algorithm behave when the size of the input grows arbitrarily large with respect to runtime and memory space usage? Answering these questions may be a little different for recursive algorithms than for some other algorithms you may have seen.

A Warm-Up, Nonrecursive Example: Powers of 2

Let us begin with the nonrecursive power-of-2 algorithm from the beginning of the chapter reprinted here:

```
function powerOf2(n)
    set product to 1
    for i from 1 to n
        set product to product * 2
    return product
```

For this algorithm, we observe a loop that starts at 1 and continues to n. This means that we have a number of steps roughly equal to n. This is a good clue that the time complexity is linear or O(n). For an n equal to 6, we could expand this procedure to the following sequence of 5 multiplications: 2 * 2 * 2 * 2 * 2 * 2. If we added 1 to n, we would have 6 multiplications for an n of 7. Actually, according to the algorithm as written, the correct sequence for an n equal to 6 would be 1 * 2 * 2 * 2 * 2 * 2 * 2. When n is 6, this sequence has exactly n multiplications counting the first multiplication by 1. This could be optimized away, but it is a good demonstration that with or without the optimization the time complexity would be O(n). As a reminder, O(n – 1) is equivalent to O(n). The time complexity O(n) is also known as **linear** time complexity.

For this algorithm, what is the space complexity? First, let us ask how many variables are used. Well, space is needed for the input parameter **n** and the **product** variable. Are any other variables needed? In a technical sense, perhaps some space would be needed to store the literal numerical values of 1 and 2, but this may depend on the specific computer architecture in use. We ignore these issues for now. That leaves the variables **n** and **product**. If the input parameter n was assigned the value of 10, how many variables would be needed? Just two variables would be needed. If n equaled 100, how many variables would be needed? Still, only two variables would be needed. This leads us to the conclusion that only a constant number of variables are needed regardless of the size of n. Therefore, the space complexity of this algorithm is O(1), also known as **constant** space complexity.

In summary, this iterative procedure takes O(n) time and uses O(1) space to arrive at the calculated result. **Note:** This analysis is a bit of a simplification. In reality, the size of the input is proportional to log(n) of the value, as the number n itself is represented in approximately $\log_2(n)$ bits. This detail is often ignored, but it is worth mentioning. The point is to build your intuition for Big-O analysis by thinking about how processes need to grow with larger inputs.

Recursive Powers of 2

Now let us consider the power-of-2 recursive algorithm from earlier and analyze its complexity.

```
function recursivePowerOf2(n)
    if n equals 0
        return 1
    else
        return 2 * recursivePowerOf2(n - 1)
```

To analyze the complexity of this algorithm, let us examine some example input values. For an n equal to 1, the algorithm begins with **recursivePowerOf2(1)**. This call evaluates **2 * recursivePowerOf2(1)**. This expression then becomes 2 * 1, which is 2. For an n equal to 3, we have the following sequence:

recursivePowerOf2(3) -> 2 * recursivePowerOf2(2)
-> 2 * 2 * recursivePowerOf2(1)
-> 2 * 2 * 2 * recursivePowerOf2(0)
-> 2 * 2 * 2 * 1

So for an **n** equal to 3, we have three multiplications. From this expansion of the calls, we can see that this process ultimately resembles the iterative process in terms of the number of steps. We could imagine that for an n equal to 6, **recursivePowerOf2(6)** would expand into 2 * 2 * 2 * 2 * 2 * 2 * 1 equivalent to the iterative process. From this, we can reason that the time complexity of this recursive algorithm is O(n). This general pattern is sometimes called a **linear recursive process**.

The time complexity of recursive algorithms can also be calculated using a recurrence relation. Suppose that we let T(n) represent the actual worst-case runtime of a recursive algorithm with respect to the input n. We can then write the time complexity for this algorithm as follows:

$$T(0) = 1$$
$$T(n) = 1 + T(n - 1).$$

This makes sense, as the time complexity is just the cost of one multiplication plus the cost of running the algorithm again with an input of n – 1 (whatever that may be). When n is 0, we must return the value 1. This return action counts as a step in the algorithm.

Now we can use some substitution techniques to solve this recurrence relation:

$$\begin{aligned} T(n) &= 1 + T(n - 1) \\ &= 1 + 1 + T(n - 2) \\ &= 2 + T(n - 2) \\ &= 2 + 1 + T(n - 3) \\ &= 3 + T(n - 3). \end{aligned}$$

Here we start to notice a pattern. If we continue this out to $T(n - n) = T(0)$, we can solve the relationship:

$$\begin{aligned} T(n) &= n + T(n - n) \\ &= n + T(0) \end{aligned}$$

=n + 1.

We see that T(n) = n + 1, which represents the worst-case time complexity of the algorithm. Therefore, the algorithm's time complexity is O(n + 1) = O(n).

Next, let us analyze the space complexity. You may try to approach this problem by checking the number of variables used in the recursive procedure that implements this algorithm. Unfortunately, recursion hides some of the memory demands due to the way procedures are implemented with the runtime stack. Though it appears that only one variable is used, every call to the recursive procedure keeps its own separate value for n. This leads to a memory demand that is proportional to the number of calls made to the recursive procedure. This behavior is illustrated in the following figure, which presents a representation of the runtime stack at a point in the execution of the algorithm. Assume that **recursivePowerOf2(3)** has been called inside main or another procedure to produce a result, and we are examining the stack when the last call to **recursivePowerOf2(0)** is made during this process.

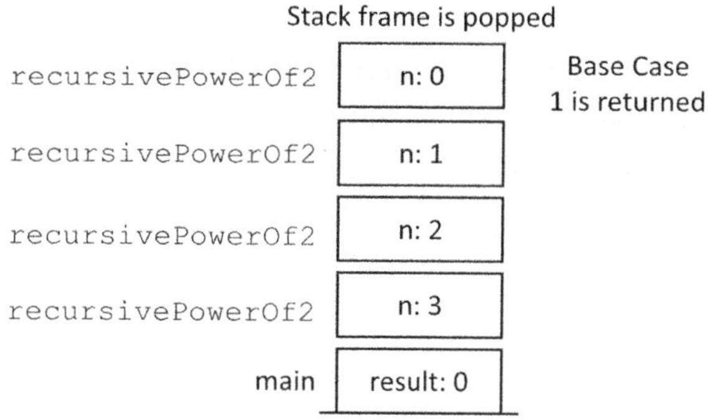

Figure 2.9

The number of variables needed to calculate **recursivePowerOf2(3)** is proportional to the size of the input. The figure showing the call stack has a frame for each call starting with 3 and going down to 0. If our input increased, our memory demand would also increase. This observation leads to the conclusion that this algorithm requires O(n) space in the worst case.

Tail-Call Optimization

In considering the two previous algorithms, we can compare them in terms of their time and space scaling behavior. The imperative **powerOf2** implementation uses a loop. This algorithm has a time complexity of O(n) and a space complexity of O(1). The **recursivePowerOf2** algorithm has a time complexity of O(n), but its space complexity is much worse at O(n). Fortunately, an optimization trick exists that allows recursive algorithms to reduce their space usage. This trick is known as tail-call optimization. If your language or compiler supports tail-call optimization, recursive algorithms can be structured to use the same amount of space as their corresponding iterative implementations. Actually, both algorithms would be considered iterative, as they iteratively update a fixed set of state variables. Let us examine a tail-call optimization example in greater detail.

To fix our **recursivePowerOf2** implementation's space complexity issues, we will first slightly modify our algorithm. We will add a state variable that will hold the running product. This serves the same purpose as the product variable in our iterative loop implementation.

```
function recursivePowerOf2(product, n)
    if n equals 0
        return product
    else
        return recursivePowerOf2(2 * product, n - 1)
```

For this algorithm to work correctly, any external call should be made with **product** set to 1. This ensures that an exponent of n equal to 0 returns 1. This means it would be a good idea to wrap this algorithm in a wrapper function to avoid unnecessary errors when a programmer mistakenly calls the algorithm without **product** set to 1. This simple wrapper function is presented below:

```
function newRecursivePowerOf2(n)
    recursivePowerOf2(1, n)
```

To understand how tail-call optimization works, let us think about what would happen on the stack when we made a call like **recursivePowerOf2(1, 3)** using our new algorithm (we will ignore the wrapper for this discussion). The stack without tail-call optimization would look like the following figure:

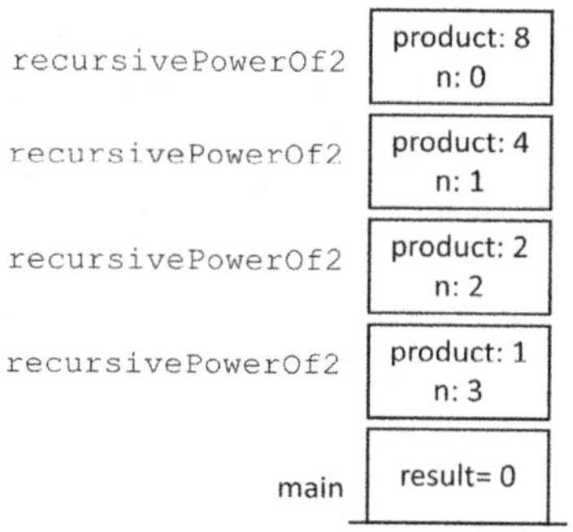

Figure 2.10

The key observation of tail-call optimization is that when the recursive call is at the tail end of the procedure (in this case, line 5), no information is needed from the current stack frame. This means that the current frame can simply be reused without consuming more memory. The following figure gives an illustration of **recursivePowerOf2(1, 3)** after the first recursive tail-call in the execution process.

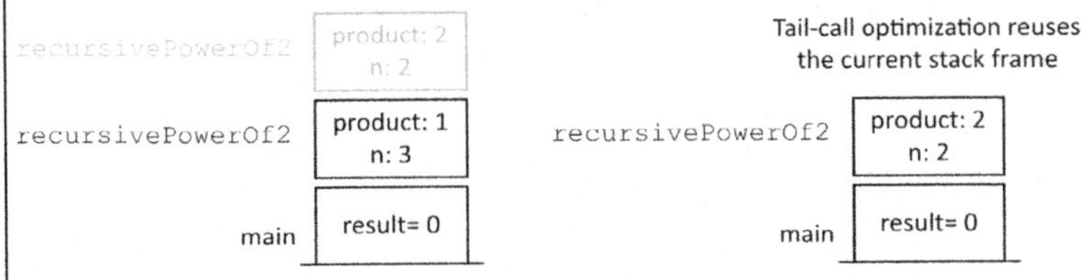

Figure 2.11

Using tail-call optimization, our new **recursivePowerOf2** algorithm has the same time complexity O(n) and space complexity O(1) as the loop-based iterative implementation. Keep in mind that tail-call optimization is a feature of either your interpreter or the compiler. You may wish to check if it is supported by your language (https://en.wikipedia.org/wiki/Tail_call#By_language).

Powers of 2 in O(log n) Time

A clever algorithm is presented in Structure and Interpretation of Computer Programs (SICP, Abelson and Sussman, 1996) that calculates powers of any base in O(log n) time. Let us examine this algorithm to help us understand some recursive algorithms that are faster—or rather, scale better—than O(n) time.

We will examine the same problem as above, calculating the nth power of 2. The key idea of this algorithm takes advantage of the fact that $2^n = (2^{n/2})^2$ when n is even and $2 * 2^{n-1}$ when n is odd. To implement this algorithm, we must first define two helper functions. One function will square an input number, and the other function will check if a number is even. We will define them here using **mod** to indicate the remainder after division.

```
function isEven(n)
    if n mod 2 equals 0
        return True
    else
        return False

function square(n)
    return n * n
```

After defining these two functions, we will consider the bases and recursive cases for which we need to account. If n is 0, the algorithm should return 1 according to the mathematical definition of any base value raised to an exponent of 0. This will be one of our cases. Next, if the n is even, we will square the result of a recursive call to the algorithm passing the value of n/2. This code is equivalent to our calculation of $(2^{n/2})^2$. Lastly, for the odd case, we will calculate 2 times the result of a recursive call that passes the value of n − 1. This code handles the odd case equivalent to $2 * 2^{n-1}$ and sets up the use of our efficient even exponent case. The algorithm is presented below:

```
function fastPowerOf2(n)
    if n equals 0
        return 1
    else if isEven(n)
        return square(fastPowerOf2(n/2))
    else
        return 2 * fastPowerOf2(n - 1)
```

Let us think about the process this algorithm uses to calculate 2^9. This process starts with a call to **fastPowerOf2(9)**. Since 9 is odd, we would multiply **2 * fastPowerOf2(8)**. This expression then expands to **2 * square(fastPowerOf2(4))**. Let us look at this in more detail.

fastPowerOf2(9)->2 * fastPowerOf2(8)
->2 * square(fastPowerOf2(4))
->2 * square(square(fastPowerOf2(2)))
->2 * square(square(square(fastPowerOf2(1))))
->2 * square(square(square(2 * fastPowerOf2(0))))
->2 * square(square(square(2 * 1)))
->2 * square(square(4))
->2 * square(16)
->2 * 256
->512

It may not be clear that this algorithm is faster than the previous **recursivePowerOf2**, which was bounded by O(n). Considering the linear calculation of 2^9 would look like this: 2 * 2 * 2 * 2 * 2 * 2 * 2 * 2 * 2 with 9 multiplications. For this algorithm, we could start thinking about the calculation that is equivalent to **2 * square(square(square(2 * 1)))**. Our first multiplication is 2 * 1. Next, the **square** procedure multiplies 2 * 2 to get 4. Then 4 is squared, and then 16 is squared for 2 more multiplications. Finally, we get 2 * 256, giving the solution of 512 after only 5 multiply operations. At least for an n of 9, **fastPowerOf2** uses fewer multiply operations. From this, we could imagine that for larger values of n, the **fastPowerOf2** algorithm would yield even more savings with fewer total operations compared to the linear algorithm.

Let us now examine the time complexity of **fastPowerOf2** using a recurrence relation. The key insight is that roughly every time the recursive algorithm is called, n is cut in half. We could model this as follows: Let us assume that there is a small constant number of operations associated with each call to the recursive procedure. We will represent this as c. Again, we will use the function T(n) to represent the worse-case number of operations for the algorithm. So for this algorithm, we have $T(n) = c + T(n/2)$. Let us write this and the base case as a recurrence relation:

$$T(0)=c$$

$$T(n)=c + T(n/2).$$

To solve this recurrence problem, we begin by making substitutions and looking for a pattern:

$$\begin{aligned} T(n) &= c + T(n/2) \\ &= c + c + T(n/4) \\ &= c + c + c + T(n/8). \end{aligned}$$

We are beginning to see a pattern, but it may not be perfectly clear. The key is in determining how many of those constant terms there will be once the expression of $T(n/x)$ becomes 1 or 0. Let us rewrite these terms in a different way:

$$T(n) = c + T(n/2^1)$$
$$= 2c + T(n/2^2)$$
$$= 3c + T(n/2^3).$$

Now we seek a pattern that is a little clearer:

$$T(n) = k * c + T(n/2^k).$$

So when will $n/2^k$ be 1?

We can solve for k in the following equation:

$$n/2^k = 1$$
$$2^k * (n/2^k) = 2^k * 1$$
$$n = 2^k.$$

Now taking the $\log_2$ of each side gives the following:

$$\log_2 n = k.$$

Now we can rewrite the original formula:

$$T(n)=\log_2 n * c + T(1).$$

According to our algorithm, with **n** equal to 1, we use the odd case giving 2 * **fastPowerOf2(0)** = 1. So we could reason that T(1) = 2c. Finally, we can remove the recursive terms and with a little rewriting arrive at the time complexity:

$$T(n)=\log_2 n * c + T(1)$$
$$=\log_2 n * c + 2c$$
$$=c * (\log_2 n + 2).$$

Therefore, our asymptotic time complexity is O(log n)

Here we have some constants and lower-order terms that lead us to a time complexity of O(log n). This means that the scaling behavior of **fastPowerOf2** is much, much better than our linear version. For reference, suppose n was set to 1,000. A linear algorithm would take around 1,000 operations, whereas an O(log n) algorithm would only take around 20 operations.

As for space complexity, this algorithm does rely on a nested call that needs information from previous executions. This requirement is due to the square function that needs the result of the recursive call to return. This implementation is not tail-recursive. In determining the space complexity, we need to think about how many nested calls need to be made before we reach a base case to signal the end of recursion. For a process like this, all these calls will consume memory by storing data on the runtime stack. As n is divided each time, we will reach a base case after O(log n) recursive calls. This means that in the worst case, the algorithm will have O(log n) stack frames taking up memory. Put simply, its space complexity is O(log n).

Exercises

1. Write a recursive function like that of **powerOf2** called "pow" that takes a **base** and **exponent** variables and computes the value of **base** raised to the **exponent** power for any integer base >= 1 and any integer exponent >= 0.
2. The sequence {0, 1, 3, 6, 10, 15, 21, 28, ... } gives the sequence of triangular numbers.

    a. Give a recursive definition of the triangular numbers (starting with n = 0).
    b. Give a recursive algorithm for calculating the nth triangular number.

3. Modify the **recMin** algorithm to create a function called **recMinIndex**. Your algorithm should return the index of the minimum value in the array rather than the minimum value itself. **Hint:** You should add another parameter to the helper function that keeps track of the minimum value's index.
4. Write an algorithm called **simplify** that prints the

simplified output of a fraction. Have the procedure accept two integers representing the numerator and denominator of a fraction. Have the algorithm print a simplified representation of the fraction. For example, **simplify(8, 16)** should print "1 / 2." Use the GCD algorithm as a subroutine in your procedure.

5.

Implement the three **powerOf2** algorithms (iterative, recursive, and **fastPowerOf2**) in your language of choice. Calculate the runtime of the different algorithms, and determine which algorithms perform best for which values of n. Using your data, create a "super" algorithm that checks the size of n and calls the most efficient algorithm depending on the value of n.

References

Abelson, Harold, and Gerald Jay Sussman. *Structure and Interpretation of Computer Programs*. The MIT Press, 1996.

3. Sorting

Learning Objectives

After reading this chapter you will...

- understand the problem of sorting a set of numbers (or letters) in a defined order.
- be able to implement a variety of well-known sorting algorithms.
- be able to evaluate the efficiency and relative advantages of different algorithms given different input cases.
- be able to analyze sorting algorithms to determine their average-case and worst-case time and space complexity.

Introduction

Applying an order to a set of objects is a common general problem in life as well as computing. You may open up your email and see that the most recent emails are at the top of your inbox. Your favorite radio station may have a top-10 ranking of all the new songs based on votes from listeners. You may be asked by a relative to put a shelf of books in alphabetical order by the authors' names. All these scenarios involve ordering or ranking based on some value.

To achieve these goals, some form of sorting algorithm must be used. A key observation is that these sorting problems rely on a specific comparison operator that imposes an ordering ("a" comes before "b" in alphabetical order, and 10 < 12 in numerical order). As a terminology note, alphabetical ordering is also known as lexical, lexicographic, or dictionary ordering. Alphabetical and numerical orderings are usually the most common orderings, but date or calendar ordering is also common.

In this chapter, we will explore several sorting algorithms. Sorting is a classic problem in computer science. These algorithms are classic not because you will often need to write sorting algorithms in your professional life. Rather, sorting offers an easy-to-understand problem with a diverse set of algorithms, making sorting algorithms an excellent starting point for the analysis of algorithms.

To begin our study, let us take a simple example sorting problem and explore a straightforward algorithm to solve it.

An Example Sorting Problem

Suppose we are given the following array of 8 values and asked to sort them in increasing order:

43	27	45	24	35	47	22	48

Figure 3.1

How might you write an algorithm to sort these values?

Our human mind could easily order these numbers from smallest to largest without much effort. What if we had 20 values? 200? We would quickly get tired and start making mistakes. For these values, the correct ordering is 22, 24, 27, 35, 43, 45, 47, 48. Give yourself some time to think about how you would solve this problem. Don't consider arrays or indexes or algorithms. Think about doing it just by looking at the numbers. Try it now.

Reflect on how you solved the problem. Did you use your fingers to mark the positions? Did you scan over all the values multiple times? Taking some time to think about your process may help you understand how a computer could solve this problem.

One simple solution would be to move the smallest value in the list to the leftmost position, then attempt to place the next smallest value in the next available position, and so on until reaching the last value in the list. This approach is called Selection Sort.

Selection Sort

Selection Sort is an excellent place to start for algorithm analysis. This sort can be constructed in a very simple way using some bottom-up design principles. We will take this approach and work our way up to a conceptually simple sorting algorithm. Before we get started, let us outline the Selection Sort algorithm. As a reminder, we will use 0-based indexing with arrays:

- Start by considering the first or 0 position of the array.
- Find the index of the smallest value in the array from position 0 to the end.
- Exchange the value in position 0 with the smallest value (using the index of the smallest value).
- Repeat the process by considering position 1 of the array (as the smallest value is now in position 0).

The algorithm works by repeatedly selecting the smallest value in the given range of the array and then placing it in its proper position along the array starting in the first position. With a little thought for our design, we can construct this algorithm in a way that greatly simplifies its logic.

Selection Sort Implementation

Let us start by creating the exchange function. Our exchange function will take an array and two indexes. It will then swap the value in the given positions within the array. For example, **exchange(array, 1, 3)** will take the value in position 1 and place it in position 3 as well as taking the value in position 3 and placing it in position 1. Let us look at what might happen by calling it on our previous array.

Here is our previous array with indexes added:

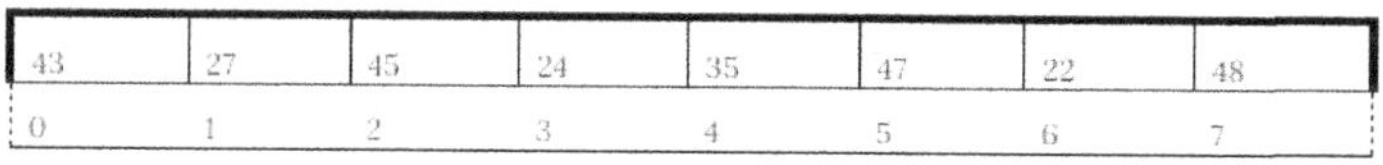

Figure 3.2

After calling **exchange(array, 1, 3)**, we get this:

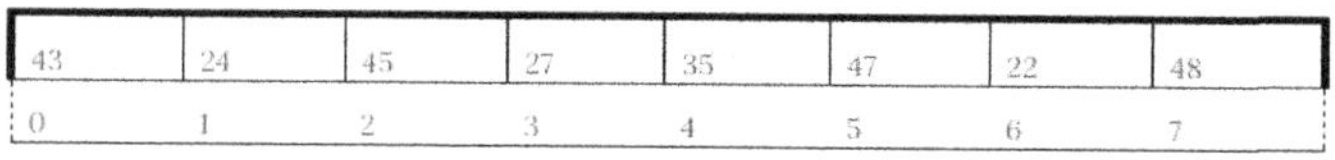

Figure 3.3

This function is the first tool we need to build Selection Sort. Let's explore one implementation of this function.

```
function exchange(array, indexA, indexB)
    set temporaryValue to array[indexA]
    set array[indexA] to array[indexB]
    set array[indexB] to temporaryValue
```

This function will do nothing if the indexes are identical. When we have separate indexes, the corresponding values in the array are exchanged. This evolves the state of the array by switching two values. In this implementation, we do not make any checks to see if an exchange could be made. It may be worth checking if the indexes are identical. It also may be worth checking if the indexes are valid (for example, between 0 and n – 1), but this exercise is left to the reader.

Now we need another tool to help us "select" the next smallest value to put in its correct order. For this task, we need something that is conceptually the same as a **findMin** function. For our algorithms, we would need to make a few modifications to the regular **findMin**. The two additions we need to make are as follows: (1) We need to get the index of the smallest value, not just its value. (2) We want to search only within a given range. The last modification will let us choose the smallest value for position 0 and then choose the second smallest value for position 1 (chosen from positions 1 to n – 1).

Let us look at one implementation for this algorithm.

```
function findMinIndex(array, start)
    set minimum to array[start]
    set minIndex to start
    set end to length(array)
    for index from start + 1 to end - 1
        if array[index] < minimum
            set minimum to array[index]
            set minIndex to index
    return minIndex
```

For this algorithm, we can set any start coordinate and find the index of the smallest value from the start to the end of the array. This simple procedure gives us a lot of power, as we will learn.

Now that we have our tools created, we can write Selection Sort. This leads to a simple implementation thanks to the design that decomposed the problem into smaller tasks.

```
function selectionSort(array)
    set end to length(array)
    for index from 0 to end - 1
        set minIndex to findMinIndex(array, index)
        exchange(array, index, minIndex)
```

We now have our first sorting algorithm. This algorithm provides a great example of how design impacts the complexity of an implementation. Combining a few simple ideas leads to a powerful new tool. This practice is sometimes called **encapsulation**. The complexity of the algorithm is encapsulated behind a few functions to provide a simple interface. Mastering this art is the key to becoming a successful computing professional. Amazing things can be built when the foundation is functional, and good design removes a lot of the difficulty of programming. Try to take this lesson to heart. Good design gives us the perspective to program in a manner that is closer to the way we think. Context that improves our ability to think about problems improves our ability to solve problems.

Selection Sort Complexity

It should be clear that the sorting of the array using Selection Sort does not use any extra space other than the original array and a few extra variables. The space complexity of Selection Sort is O(n).

Analyzing the time complexity of Selection Sort is a little trickier. We may already know that the complexity of finding the minimum value from an array of size n is O(n) because we cannot avoid checking every value in the array. We might reason that there is a loop that goes from 1 to n in the algorithm, and our findMinIndex should also be O(n). This idea leads us to think that calling an O(n) function n times would lead to $O(n^2)$. Is this correct? How can we be sure? Toward the end of the algorithm's execution, we are only looking for the minimum value's index from among 3, 2, or 1 values. This seems close to O(1) or constant time. Calling an O(1) function n times would lead to O(n), right? Practicing this type of reasoning and asking these questions will help develop your algorithm analysis skills. These are both reasonable arguments, and they have helped establish a bound for our algorithm's complexity. It would be safe to assume that the actual runtime is somewhere between O(n) and $O(n^2)$. Let us try to tackle this question more rigorously.

When our algorithm begins, nothing is in sorted order (assume a random ordering). Our index from line 3 of selectionSort starts at 0. Next, findMinIndex searches all n elements from 0 to n – 1. Then we have the smallest value in position 0, and index becomes 1. With index 1, findMinIndex searches n – 1 values from 1 to n – 1. This continues until index becomes n – 1 and the algorithm finishes with all values sorted.

We have the following pattern:

With index at 0 n comparison operations are performed by findMinIndex.

With index at 1 n – 1 comparison operations are performed by findMinIndex.

With index at 2 n − 2 comparison operations are performed by findMinIndex.

...

With index at n − 2 2 comparison operations are performed by findMinIndex.

With index at n − 1 1 comparison operation is performed by findMinIndex.

Our runtime is represented by the sum of all these operations. We could rewrite this in terms of the sum over the number of comparison operations at each step:

$$n + (n - 1) + (n - 2) + \ldots 3 + 2 + 1.$$

Can we rewrite this sequence as a function in terms of n to give the true runtime? One way to solve this sequence is as follows:

$$\text{Let } S = n + (n - 1) + (n - 2) + \ldots 3 + 2 + 1.$$

Multiplying S by 2 gives

$$2 * S=[n + (n - 1) + (n - 2) + \ldots + 3 + 2 + 1] + [n + (n - 1) + (n - 2) + \ldots + 3 + 2 + 1].$$

We can rearrange the right-hand side to highlight a useful pattern:

$$2 * S=[n + (n - 1) + (n - 2) + \ldots + 3 + 2 + 1] + [1 + 2 + 3 + \ldots + (n - 2) + (n - 1) + n].$$

We notice that lining them up with one sequence reversed leads to n terms of n + 1:

$$2 * S=(n + 1) + (n + 1) + (n + 1) + \ldots + (n + 1) + (n + 1) + (n + 1) =n * (n + 1).$$

Now we can divide by two to get an exact function for this summation sequence, which is also known as a variant of the arithmetic series:

$$S=[n * (n + 1)]/2.$$

To view it in polynomial terms, we can distribute the n term and move the fraction:

$$S=(½)*[n^2 + n]$$
$$=(½)\ n^2 + (½)\ n.$$

In Big-O terms, the time complexity of Selection Sort is $O(n^2)$. This is also known as **quadratic** time.

Insertion Sort

Insertion Sort is another classic sorting algorithm. Insertion Sort orders values using a process like organizing books on a bookshelf starting from left to right. Consider the following shelf of unorganized books:

Figure 3.4

Currently, the books are not organized alphabetically. Insertion Sort starts by considering the first book as sorted and placing an imaginary separator between the sorted and unsorted books. The algorithm then considers the first book in the unsorted portion.

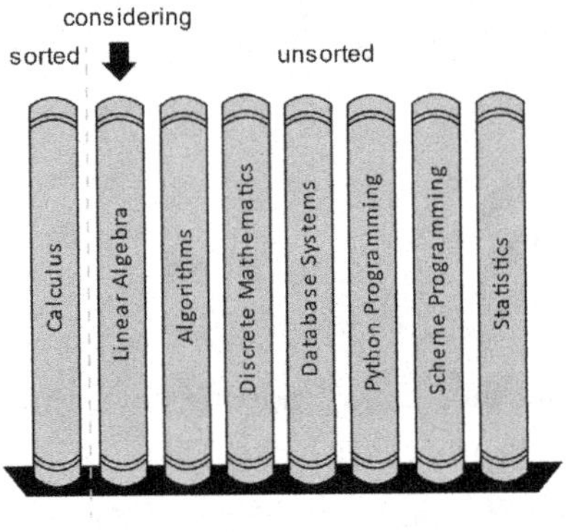

Figure 3.5

According to the image, the algorithm is now considering the book called Linear Algebra. The algorithm will now try to place this new book into its proper position in the sorted section of the bookshelf. The letter "C" comes before "L," so the book should be placed to the right of the Calculus book, and the algorithm will consider the next book. This state is shown in the following image:

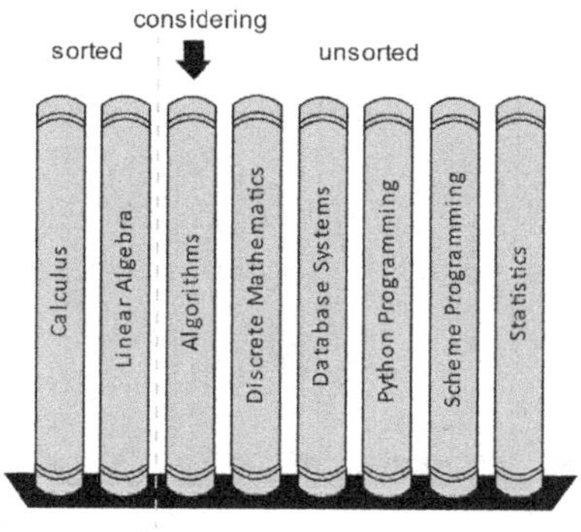

Figure 3.6

Now we consider the Algorithms book. In this case, the Algorithms book should come before both the Calculus and Linear Algebra books, but there is no room on the shelf to just place it there. We must make room by moving the other books over.

The actual process used by the algorithm considers the book immediately to the left of the book under consideration. In this case, the Linear Algebra book should come after the Algorithms book, so the Linear Algebra book is moved over one position to the right. Next, the Calculus book should come after the Algorithms book, and the Calculus book is moved one position to the right. Now there are no other books to reorder, so the Algorithms book is placed in the correct position on the shelf. This situation is highlighted in the following image:

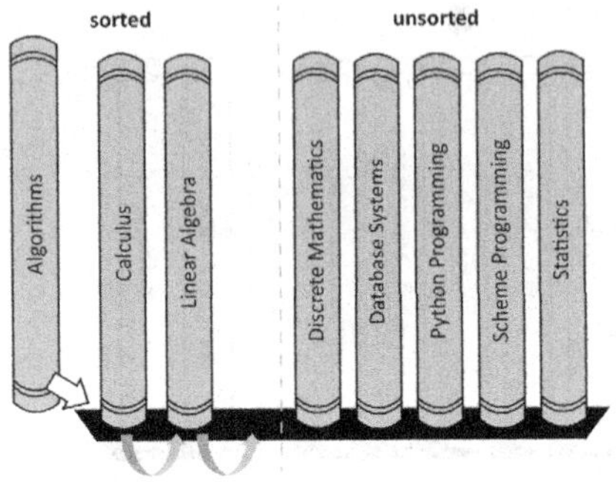

Figure 3.7

After adjusting the position of these books, we have the state displayed below:

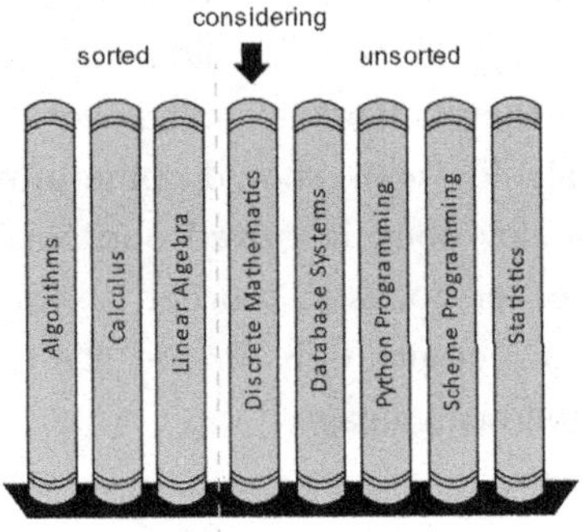

Figure 3.8

Now we consider the textbook on Discrete Mathematics. First, we examine the Linear Algebra book to its left. The Linear Algebra book should go after the Discrete Mathematics book, and it is moved to the right by one position. Now examining the Calculus book informs us that no other sorted books should be after the Discrete Mathematics book. We will now place the Discrete Mathematics book in its correct place after the Calculus book. The Algorithms book at the far left is not even examined. This process is illustrated in the figure below:

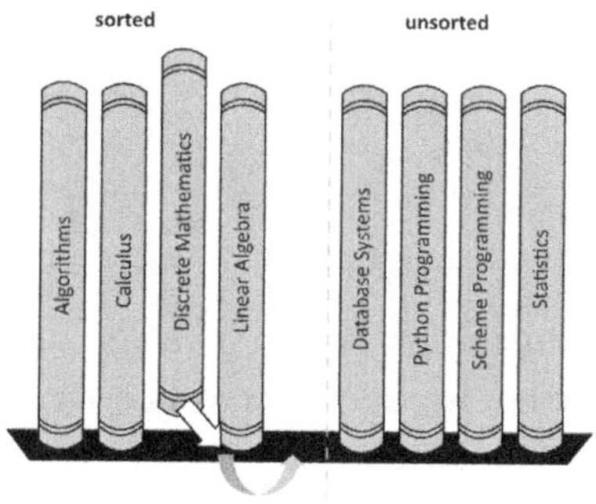

Figure 3.9

Once the Discrete Mathematics book is placed in its correct position, the process begins again, considering the next book as shown below:

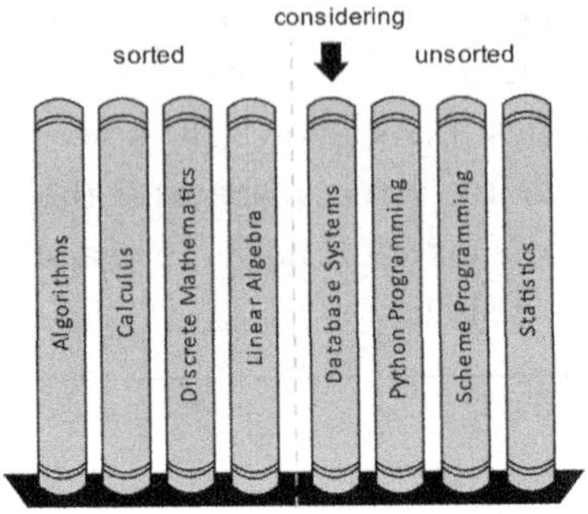

Figure 3.10

This brief illustration should give you an idea of how the Insertion Sort algorithm works. We consider a particular book and then "insert" it into the correct sorted position by moving books that come after it to the right by one position. The sorted portion will grow as we consider each remaining book in the unsorted portion. Finally, the unsorted section of the bookcase will be empty, and all the books will be sorted properly. This example also illustrates that there are other types of ordering. Numerical ordering and alphabetical ordering are probably the most common orderings you will encounter. Date and time ordering are also common, and sorting algorithms will work just as well with these as with other types of orderings.

Insertion Sort Implementation

For our implementation of Insertion Sort, we will only consider arrays of integers as with Selection Sort. Specifically, we will assume that the values of positions in the array are comparable and will lead to the correct ordering. The process will work equally well with alphabetical characters as with numbers provided the relational operators are defined for these and other orderable types. We will examine a way to make comparisons more flexible later in the chapter.

```
function insertionSort(array)
    set end to length(array)
    set endSorted to 1
    while endSorted < end
        set currentValue to array[endSorted]
        set index to endSorted
        while index > 0 and currentValue  < array[index - 1]
            # move values to the right
            set array[index] to array[index - 1]
            set index to index - 1
        # after moving items, place the current item
        set array[index] to currentValue
        set endSorted to endSorted + 1
```

This algorithm relies on careful manipulation of array indexes. Manipulation of array indexes often leads to errors as humans are rarely careful. As always, you are encouraged to test your algorithms using different types of data. Let us test our implementation using the array from before. We will do this by using a trace of an algorithm. A **trace** is just a way to write out or visualize the sequence of steps in an algorithm. Consider the following array with indexes:

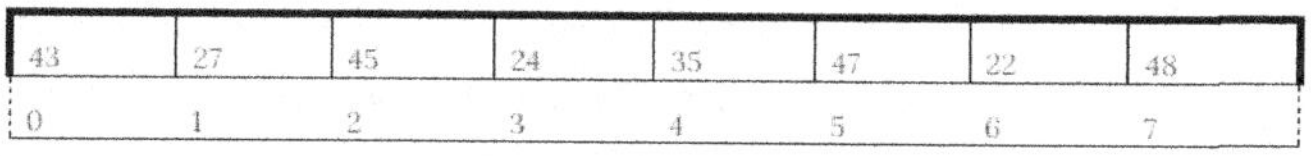

Figure 3.11

The algorithm will start as follows with **endSorted** set to

1 and end set to 8. Entering the body of the while-loop, we set **currentValue** to the value 27, and index is set to **endSorted**, which is 1. The image below gives an illustration of this scenario:

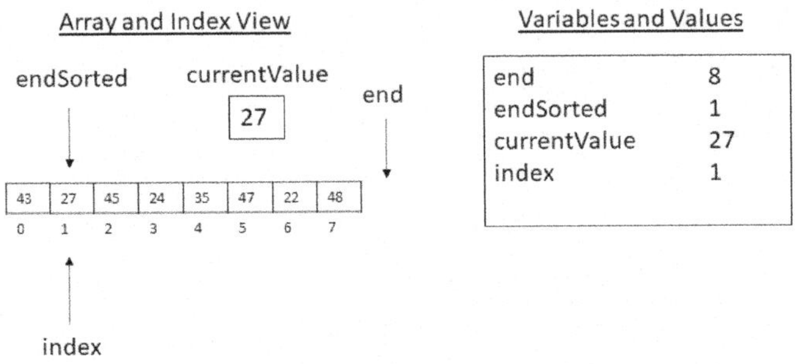

Figure 3.12

We now consider the inner loop of the algorithm. The compound condition of **index > 0 and currentValue < array[index – 1]** is under consideration. Index is greater than 0, check! This part is true. The value in **array[index – 1]** is the value at **array[1 – 1]** or **array[0]**. This value is 43. Now we consider 27 < 43, which is true. This makes the compound condition of the while-loop true, so we enter the loop. This executes **set array[index] to array[index – 1]**, which copies 43 into position **array[index]** or **array[1]**. Next, the last command is executed to update **index** to **index – 1** or 0. This gives us the following state:

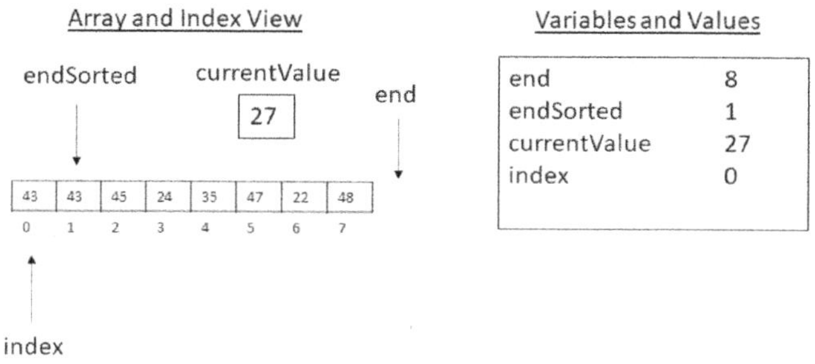

Figure 3.13

You may be thinking, "What about 27? Is it lost?" No, the value of 27 was saved in the **currentValue** variable. We now loop and check **index > 0** and **currentValue < array[index – 1]** again. This time **index > 0** fails with **index** equal to 0. The algorithm then executes **set array[index] to currentValue**. This operation places 27 into position 0, providing the correct order. Finally, the **endSorted** variable is increased by 1 and now points to the next value to consider. At the start of the next loop, after **currentValue** and **index** are set, we have the state of execution presented below:

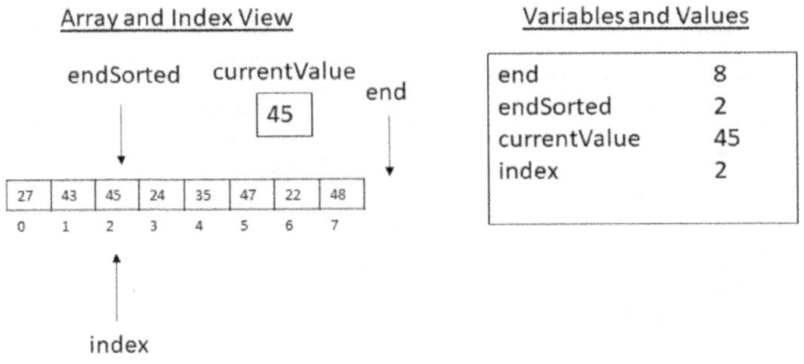

Figure 3.14

Next, the algorithm would check the condition of the inner loop. Here **index** is greater than 0, but **currentValue** is not less than **array[index - 1]**, as 45 is not less than 43. Moving over the loop, **currentValue** is placed back into its original position (line 12), and the indexes are updated at the end of the loop as usual. Moving to the next value to consider, before line 7 we now have the scenarios presented below:

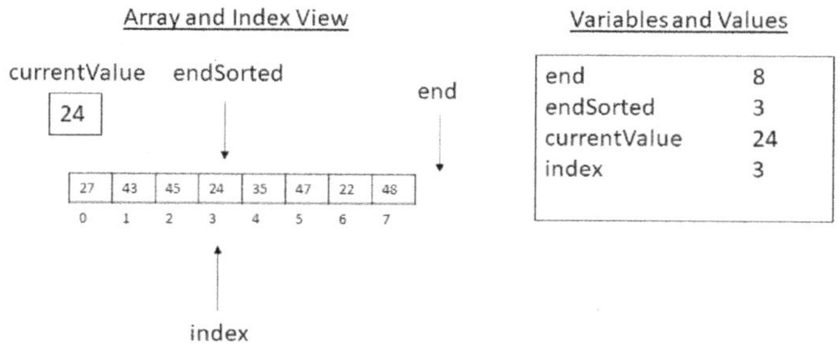

Figure 3.15

From this image, we can imagine what would happen next. The value 24 is smaller than all these values. Let us trace how the algorithm would proceed. As 24 is less than 45, the body will execute copying 45 to the right and updating the index.

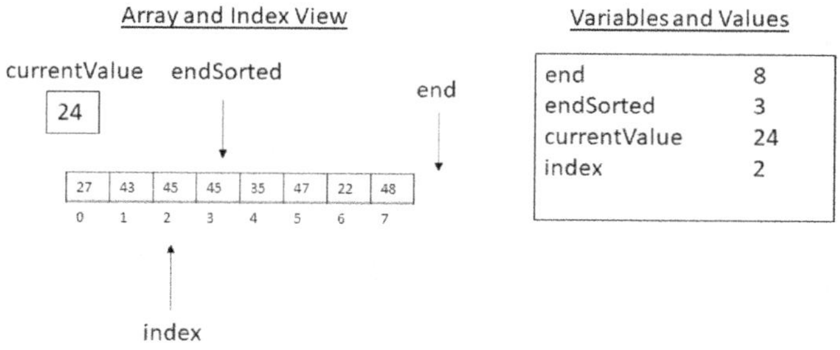

Figure 3.16

In the above image, we notice that 24 is less than 43. Therefore, we copy and update our indexes. This gives the following successive states of execution:

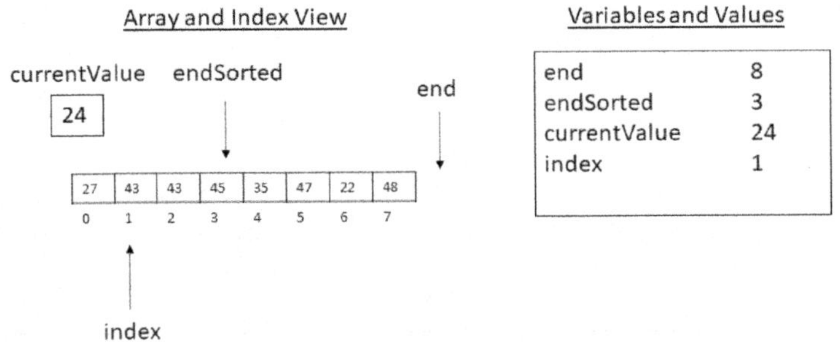

Figure 3.17

Then...

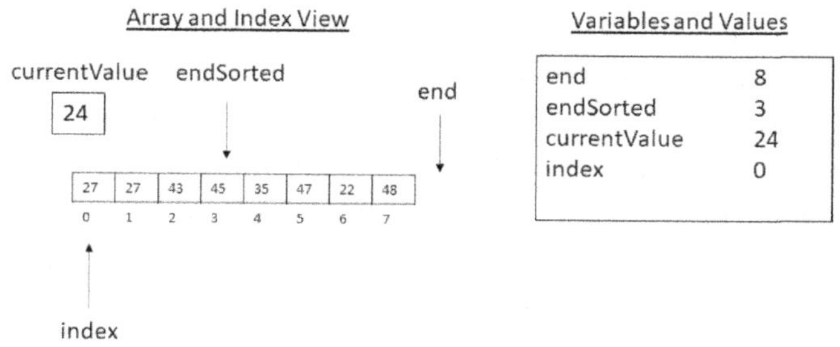

Figure 3.18

Now with index 0, the inner loop's body will not execute. We will copy the **currentValue** into the **index** position and begin the outer loop's execution again. At the start of the next inner loop, we have the scene below:

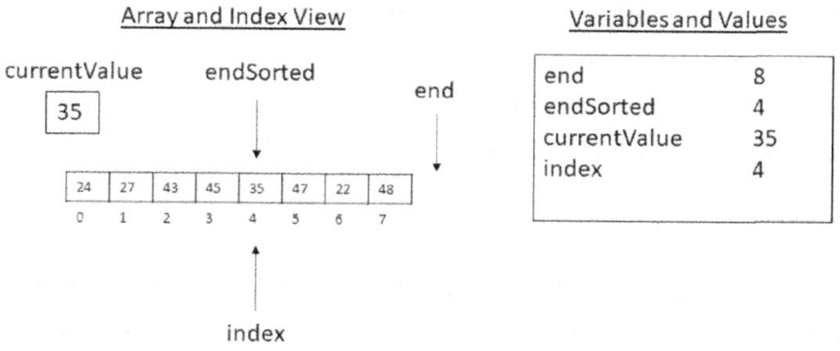

Figure 3.19

These illustrations give you an idea of the execution of the Insertion Sort. These drawings are sometimes called traces. Creating algorithm tracing will give a good idea of how your algorithm is working and will also help you understand if your algorithm is correct or not. From these diagrams, we can infer that the **endSorted** value will grow to reach the end of the array and all the values will eventually be properly sorted. You should attempt to complete the rest of this trace for practice.

Insertion Sort Complexity

From this example execution, you may have noticed that sometimes Insertion Sort does a lot of work, but other times it seems that very little needs to be done. This observation allows us to consider a different way to analyze algorithms—namely, the best-case time complexity. Before we address this question, let us analyze the worst-case space complexity and the worst-case time complexity of Insertion Sort.

The space complexity of Insertion Sort should be easy to determine. We only need space for the array itself and a few other index- and value-storage variables. This means that our memory usage is O(n) for the array and a small constant number of other values where $c \ll n$ (a constant much smaller than n). This means our memory usage is bounded by $n + c$, and we have an O(n) space complexity for Insertion Sort. The memory usage of Insertion Sort takes O(n) for the array itself and O(1) for the other necessary variables. This O(1) memory cost for the indexes and other values is sometimes referred to as the **auxiliary memory**. It is the extra memory needed for the algorithm to function in addition to the storage cost of the array values themselves. This auxiliary memory could be freed after the algorithm completes while keeping the sorted array intact. We will revisit auxiliary memory later in the chapter.

When considering the time complexity, we are generally interested in the **worst-case** scenario. When we talk about a "case," we mean a particular instance of the problem that has some special features that impact the algorithm's performance. What special features of the way our values are organized might lead to a good or bad case for our algorithm? In what situation would we encounter the absolute largest number of operations? As we observed in our trace, the value 24 resulted in a lot of comparisons and move operations. We continued to check each value and move all the values greater than 24 one space to the right. In contrast, value 45 was nearly in the right place. For 45, we only "moved" it back into the same place from which it came. Take a moment to think about what these observations might mean for our worst-case and best-case analysis.

Let us think about the case of 24 first. Why did 24 require so many operations? Well, it was smaller than all the values that came before it. In a sense, it was "maximally out of place." Suppose the value 23 came next in the **endSorted** position. This would require us to move all the other values, including 24, over again to make room for 23 in the first position. What if 22 came next? There may be a pattern developing. We considered some values followed by 24, 23, then 22. These values would lead to a lot of work each time. The original starting order of our previous array was 43, 27, 45, 24, 35, 47, 22, 48. For Insertion Sort, what would be the worst starting order? Or put another way, which starting order would lead to the absolute highest number of comparison and move operations? Take a moment to think about it.

When sorting in increasing order, the worst scenario or Insertion Sort would be an array where all values are in decreasing order. This means that every value being considered for placement is “maximally out of place.” Consider the case where our values were ordered as 48, 47, 45, 43, 35, 27, 24, 22, and we needed to place them in increasing order. The positions of 48 and 47 must be swapped. Next, 45 must move to position 0. Next, 43 moves to position 0 resulting in 3 comparisons and 3 move (or assignment) operations. Then, 35 results in 4 comparisons and 4 move operations to take its place at the front. This process continues for smaller and smaller values that need to be moved all the way to the front of the array.

From this pattern, we see that for this worst-case scenario, the first value considered takes 1 comparison and 1 move operation. The second value requires 2 comparisons and 2 moves. The third takes 3 comparisons and 3 moves, and so on. As the total runtime is the sum of the operations for all the values, we see that a function for the worst-case runtime would look like the following equation. The 2 accounts for an equal number of comparisons and moves:

$$T(n)=2 * 1 + 2 * 2 + 2 * 3 + 2 * 4 + \ldots 2 * (n - 1).$$

This can be rewritten as

$$T(n)=2 * (1 + 2 + 3 + 4 \ldots n - 1).$$

We see that this function has a growing sum as we saw with Selection Sort. We can substitute this value back into the time equation, T(n):

$$T(n)=2 * \{(½)[n*(n - 1)]\}.$$

The 2 and ½ cancel, leaving

$$T(n)=n*(n - 1).$$

Viewing this as a polynomial, we have

$$T(n)=n^2 - n.$$

This means that our worst-case time complexity is $O(n^2)$. This is the same as Selection Sort's worst-case time complexity.

Best-Case Time Complexity Analysis of Insertion Sort and Selection Sort

Now that we have seen the worst-case scenario, try to imagine the **best-case** scenario. What feature would that best-case problem instance have for Insertion Sort? In our example trace, we noticed that the value 45 saw 1 comparison and 1 "move," which simply set 45 back in the same place. The key observation is that 45 was already in its correct position relative to the other values in the sorted portion of the array. Specifically, 45 is larger than the other values in the sorted portion, meaning it is "already sorted." Suppose we next considered 46. Well, 46 would be larger than 45, which is already larger than the other previous values. This means 46 is already sorted as well, resulting in 1 additional comparison and 1 additional move operation. We now know that the best-case scenario for Insertion Sort is a correctly sorted array.

For our example array, this would be 22, 24, 27, 35, 43, 45, 47, 48. Think about how Insertion Sort would proceed with this array. We first consider 24 with respect to 22. This gives 1 comparison and 1 move operation. Next, we consider 27, adding 1 comparison and 1 move, and so on until we reach 48 at the end of the array. Following this pattern, 2 operations are needed for each of the n – 1 values to the right of 22 in the array. Therefore, we have 2*(n – 1) operations leading to a bound of O(n) operations for the best-case scenario. This means that when the array is already sorted, Insertion Sort will execute in O(n) time. This could be a significant cost savings compared to the $O(n^2)$ case.

The fact that Insertion Sort has a best-case time complexity of O(n) and a worst-case time complexity of $O(n^2)$ may be hard to interpret. To better understand these features, let us consider the best- and worst-case time complexity of Selection Sort. Suppose that we attempt to sort an array with Selection Sort, and that array is already sorted in increasing order. Selection Sort will begin by finding the minimum value in the array and placing it in the first position. The first value is already in its correct position, but Selection Sort still performs n comparisons and 1 move. The next value is considered, and n – 1 comparisons are executed. This progression leads to another variation of the arithmetic series (n + n – 1 + n – 2 ...) leading to $O(n^2)$ time complexity. Suppose now that we have the opposite scenario, where the array is sorted in descending order. Selection Sort performs the same. It searches for the minimum and moves it to the first position. Then it searches for the second smallest value, moves it to position 1, and continues with the remaining values. This again leads to $O(n^2)$ time complexity.

We have now considered the already-sorted array and the reverse-order-sorted array, and both cases led to $O(n^2)$ time complexity.

Regardless of the orderings of the input array, Selection Sort always takes $O(n^2)$ operations. Depending on the input configuration, Insertion Sort may take $O(n^2)$ operations, but in other cases, the time complexity may be closer to $O(n)$. This gives Insertion Sort a definite advantage over Selection Sort in terms of time complexity. You may rightly ask, "How big is this advantage?" Constant factors can be large, after all. The answer is "It depends." We may wish to ask, "How likely are we to encounter our best-case scenario?" This question may only be answered by making some assumptions about how the algorithms will be used or assumptions about the types of value sets that we will be sorting. Is it likely we will encounter data sets that are nearly sorted? Would it be more likely that the values are in a roughly random order? The answers to these questions will be highly context-dependent. For now, we will only highlight that Insertion Sort has a better best-case time complexity bound than Selection Sort.

Merge Sort

Now we have seen two interesting sorting algorithms for arrays, and we have had a fair amount of experience analyzing the time and space complexity of these algorithms. Both Selection Sort and Insertion Sort have a worst-case time complexity bound of $O(n^2)$. In computer science terms, we would say that they are both equivalent in terms of runtime. "But wait! You said Insertion Sort was faster!" Yes, Insertion Sort may occasionally perform better, but it turns out that even the average-case runtime is $O(n^2)$. In general, their growth functions are both bounded by $O(n^2)$. This fact means that when n is large enough, any minor differences between their actual runtime efficiency will become negligible with respect to the overall time: Graduate Student: "This algorithm is more efficient! It will only take 98 years to complete compared to the 99 years of the other algorithm." Professor: "I would like to have the problem solved before I pass away. Preferably, before I retire." This is an extreme example. Often, minor improvements to an algorithm can make a very real impact, especially on real-time systems with small input sizes. On the other hand, reducing the runtime bound by more than a constant factor can have a drastic impact on performance. In this section, we will present Merge Sort, an algorithm that greatly improves on the runtime efficiency of Insertion Sort and Selection Sort.

Description of Merge Sort

Merge Sort uses a recursive strategy to sort a collection of numbers. In simple terms, the algorithm takes two already sorted lists and merges them into one final sorted list. The general strategy of dividing the work into subproblems is sometimes called "divide and conquer." The algorithm can be specified with a brief description.

To Merge Sort a list, do the following:

- Recursively Merge Sort the first half of the list.
- Recursively Merge Sort the second half of the list.
- Merge the two sorted halves of the list into one list.

Before we investigate the implementation, let us visualize how this might work with our previous array. Suppose we have the following values:

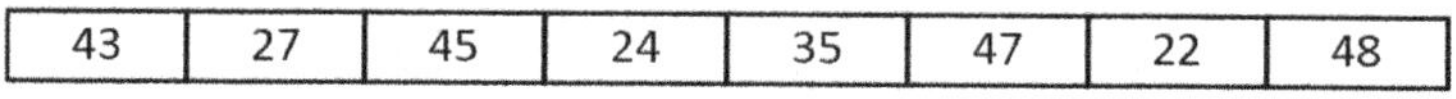

Figure 3.20

Merge Sort would first split the array in half and then make a recursive call on the two subarrays. This would in turn split repeatedly until each array is only a single element. For this example, the process would look something like the image below.

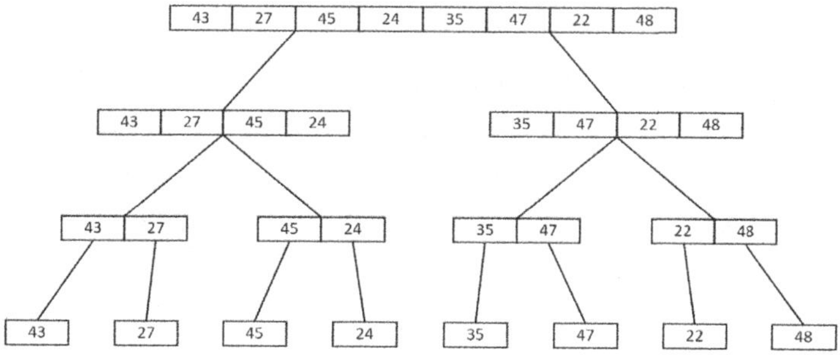

Figure 3.21

Now the algorithm would begin merging each of the individual values into sets of two, then two sets of two into a sorted list of four, and so on. This is illustrated below:

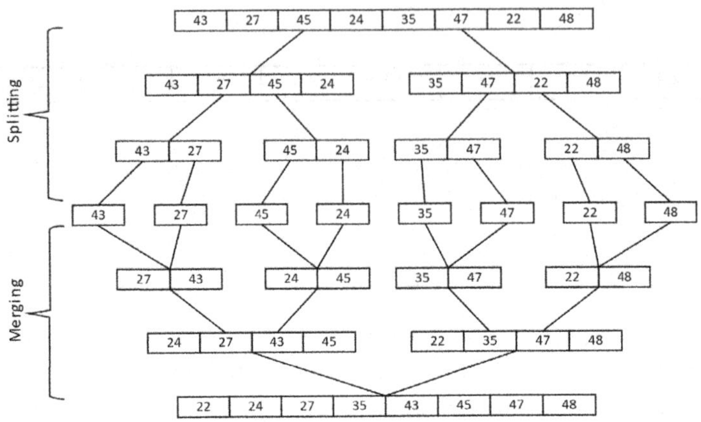

Figure 3.22

The result would be a correctly sorted list. This simple strategy leads to an efficient algorithm, as we will see.

There is one part of the process we have not discussed: the merge process. Let us explore merging at one of the intermediate levels. The example below shows one of the merge steps. The top portion shows conceptually what happens. The bottom portion shows that specific items are moved into specific positions in the new sorted array.

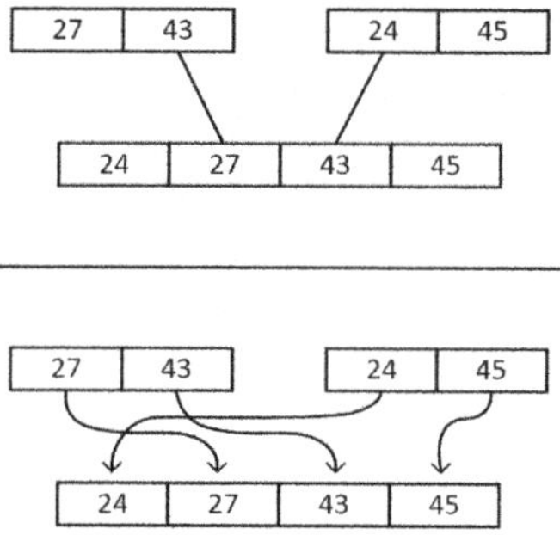

Figure 3.23

To implement Merge Sort, the merge function will be an important part of the algorithm. Let us explore merging with another diagram that more closely resembles our eventual design. At this intermediate step, the data of interest is part of a larger array. These data are composed of two sorted sublists. There are some specific indexes we want to know about. Specifically, they are the start of the first sublist, the end of the first sublist, and the final position of the data representing the end of the data to be merged.

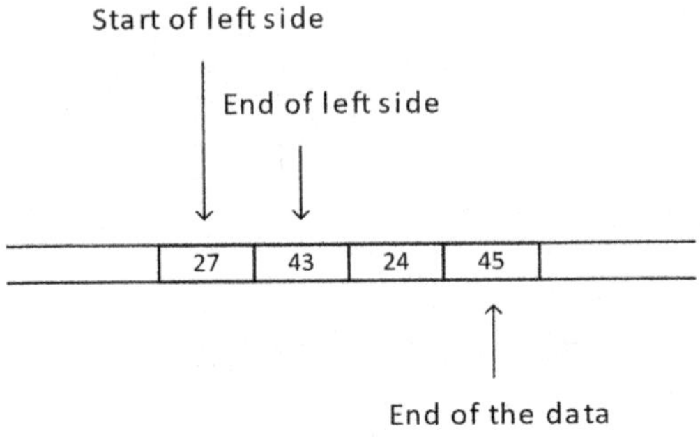

Figure 3.24

Our strategy for merging will be to copy the two sublist sets of values into two temporary storage arrays and then merge them back into the original array. This process is illustrated below:

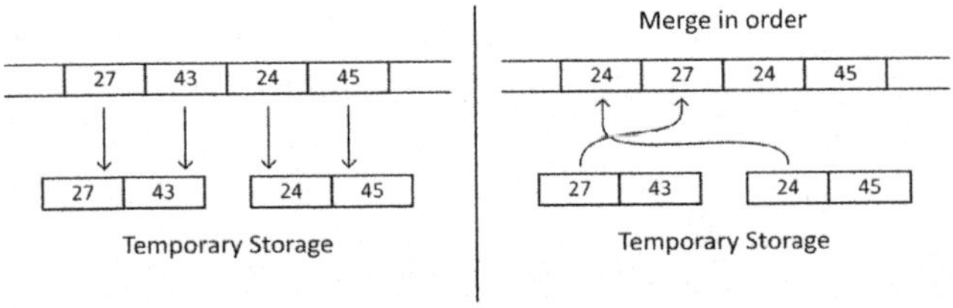

Figure 3.25

Once the merge process completes, the temporary storage may be freed. This requires some temporary memory usage, but this will be returned to the system once the algorithm completes.

Using this approach simplifies the code and makes sure the data are sorted and put back into the original array.

These are the key ideas behind Merge Sort. Now we will examine the implementation.

Merge Sort Implementation

The implementation of algorithms can be tricky. Again, we will use the approach of creating several functions that work together to solve our problem. This approach has several advantages. Mainly, we want to reduce our cognitive load so that thinking about the algorithm becomes easier. We will make fewer errors if we focus on specific components of the implementation in sequence rather than trying to coordinate multiple different ideas in our minds.

We will start with writing the **merge** function. This function will accept the array and three index values (start of left half, end of left half, and end of data). Notice that for this merge implementation, we use end to refer to the last valid position of the subarray. This use of end is a little different from how we have used it before. Before, we had end specify a value *beyond* the array (one position beyond the last valid index). Here end will specify the *last valid index*. The merge function proceeds by copying the data into temporary storage arrays and then merging them back into the original array.

For the copy step, we extend the storage arrays to hold one extra value. This value is known as a **sentinel**. The sentinel will be a special value representing the highest possible value (or lowest for other orderings). Most programming languages support this as either a **MAX** or an **Infinity** construct. Any ordinary value compared to **MAX** or **Infinity** will be less than the sentinel (or greater than for **MIN** or **−Infinity** with other orderings). In our code, we will use **MAX** to represent our sentinel construct.

The merge function is presented below:

```
function merge(array, startLeft, endLeft, end)
    set sizeLeft to endLeft - startLeft + 1
    set sizeRight to end - endLeft
    set startRight to endLeft + 1
    # create temporary arrays with one extra element to store MAX
    set arrayLeft to new array of size (sizeLeft + 1)
    set arrayRight to new array of size (sizeRight + 1)

    # copy values to temporary storage
    for index from 0 to sizeLeft - 1
        set arrayLeft[index] to array[startLeft + index]
    set arrayLeft[sizeLeft] to MAX

    for index from 0 to sizeRight - 1
        set arrayRight[index] to array[startRight + index]
    set arrayRight[sizeRight] to MAX

    # merge the temporary arrays into the original array
    set indexLeft to 0
    set indexRight to 0
    for index from startLeft to end
        if arrayLeft[indexLeft] <= arrayRight[indexRight]
            set array[index] to arrayLeft[indexLeft]
            set indexLeft to indexLeft + 1
        else
            set array[index] to arrayRight[indexRight]
            set indexRight to indexRight + 1
```

This function will take a split segment of an array, identified by indexes, and merge the values of the left and right halves in order. This will place their values into their proper sorted order in the original array. **Merge** is the most complex function that we will write for Merge Sort. This function is complex in a general sense because it relies on the careful manipulation of indexes. This is a very error-prone process that leads to many off-by-one errors. If you forget to add 1 or subtract 1 in a specific place, your algorithm may be completely broken. One advantage of a modular design is that these functions can be tested independently. At this time, we may wish to create some tests for the **merge** function. This is a good practice, but it is outside of the scope of this text. You are encouraged to write a test of your newly created function with a simple example such as 27, 43, 24, 45 from the diagrams above.

Now that the **merge** implementation is complete, we will move on to writing the recursive function that will complete Merge Sort. Here we will use the mathematical "floor" function. This is equivalent to integer division in C-like languages or truncation in languages with fixed-size integers.

```
function mergeSort(array, start, end)
    if start < end
        set endLeft to floor((start + end)/2)
        mergeSort(array, start, endLeft)
        mergeSort(array, endLeft + 1, end)
        merge(array, start, endLeft, end)
```

From this implementation, we see that the recursion continues while the **start** index is less than the **end** index. Recursion ends once the **start** and **end** index have the same value. In other words, our process will continue splitting and splitting the data until it reaches a level of a single-array element. At this point, recursion ends, a single value is sorted by default, and the merging process can begin. This process continues until the final two halves are sorted and merged into their new positions within the starting array, completing the algorithm.

Once more, we have a recursive algorithm requiring some starting values. In cases like these, we should use a wrapper to provide a more user-friendly interface. A wrapper can be constructed as follows:

```
function mergeSortWrapper(array)
    set size to length(array)
    mergeSort(array, 0, size - 1)
```

Merge Sort Complexity

At the beginning of the Merge Sort section, we stated that Merge Sort is indeed faster than Selection Sort and Insertion Sort in terms of worst-case runtime complexity. We will look at how this is possible.

We may begin by trying to figure out the complexity of the **merge** function. Suppose we have a stretch of n values that need to be merged. They are copied into two storage arrays of size roughly n/2 and then merged back into the array. We could reason that it takes n/2 individual copies for each half of the array. Then another n copies back into the original array. This gives n/2 + n/2 + n total copy operations, giving 2n. This would be O(n) or linear time.

Now that we know merging is O(n) we can start to think about Merge Sort. Thinking about the top-level case at the start of the algorithm, we can set up a function for the time cost of Merge Sort:

$$T(n)=2*T(n/2) + c*n.$$

This captures the cost of Merge Sorting the two halves of the array and the merge cost, which we determined would be O(n) or n times some constant c. Substituting this equation into itself for T(n/2) gives the following:

$$T(n)=2*[2 * T(n/4) + c*n/2] + c*n.$$

Cleaning things up a little gives the following sequence:

$$T(n)=2*[2 * T(n/4) + c*(n/2)] + c*n$$
$$=4 * T(n/4) + 2*c*(n/2) + c*n$$
$$=4 * T(n/4) + c*n + c*n.$$

As this pattern continues, we will get more and more c*n

terms. Instead of continuing the recurrence, we will instead draw a diagram to show how many of these we can expect.

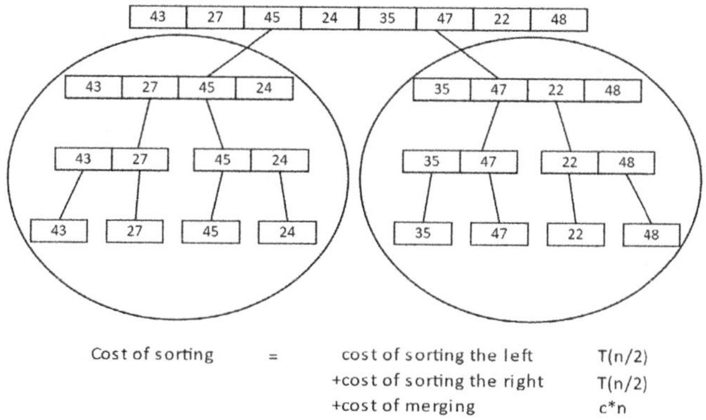

Figure 3.26

Expanding the cost of sorting the two halves, we get the next diagram.

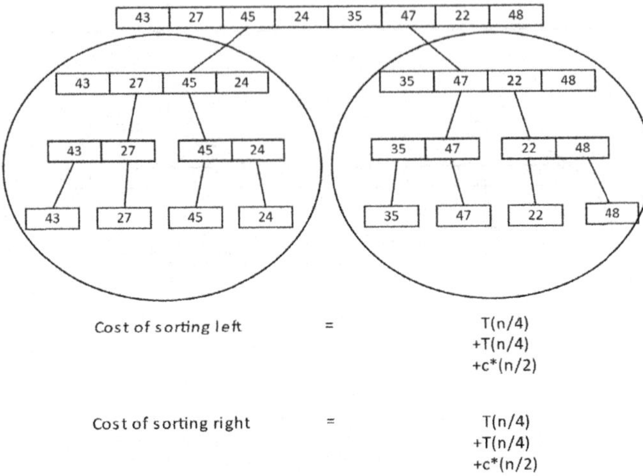

Figure 3.27

As the process continues, the c*n terms start to add up.

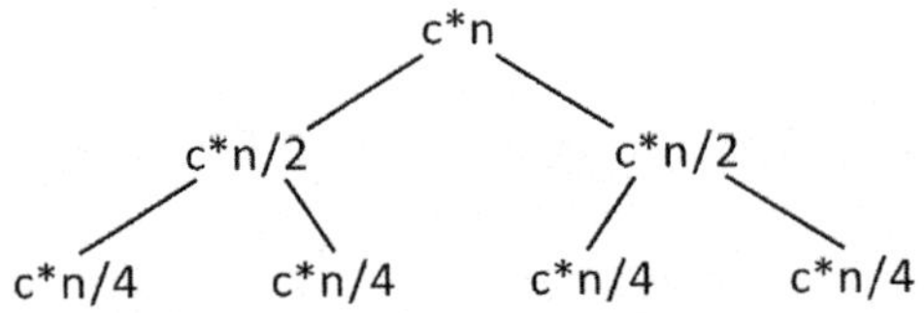

Figure 3.28

To determine the complexity though, we need to know how many of these terms to expect. The number of c*n terms is related to the depth of the recursion. We need to know how many times to split the array before arriving at the case where the start index is equal to the end index. In other words, how many times can we split before we reach the single-element level of the array? We just need to solve $1 = n/2^k$ for the value k. Setting k to $\log_2 n$ solves this equation. Therefore, we have $\log_2 n$ occurrences of the c*n terms. If we assume that n is a power of 2, the overall time cost gives us $T(n) = n*T(1) + \log_2 n * c * n$. For T(1), our wrapper function would make one check and return. We can safely assume that T(1) = c, a small constant. We could then write it as $T(n) = n*c + \log_2 n * c * n$, or $c*(n * \log_2 n + n)$. In Big-O terms, we arrive at $O(n \log_2 n)$. A consequence of Big-O is that constants are ignored. Logs in any base can be related to each other by a constant factor ($\log_2(n) = \log_3(n)/\log_3(2)$, and note that $1/\log_3(2)$ is a constant), so the base is usually dropped in computer science discussions. We can now state the proper worst-case runtime complexity of Merge Sort is O(n log n). It may not be obvious, but this improvement leads to a fundamental improvement over $O(n^2)$. For example, at n = 100, n^2 = 10000, but $n*\log_2 n$ is approximately 665, which is less than a tenth of the n^2 value. Merge Sort guarantees a runtime bounded by O(n log n), as the best case and worst case are equivalent (much like Selection Sort).

For space complexity, we will need at least enough memory as we have elements of the array. So we need at least O(n) space. Remember that we also needed some temporary storage for copying the subarrays before merging them again. We need the most memory for the case near the end of the algorithm. At this stage, we have two n/2-sized storage arrays in addition to the original array. This leads us to space for the original n values and another n value's worth of storage for the temporary values. That gives 2*n near the end of the calculation, so overall memory usage seems to be O(n). This is not the whole story though.

Since we are using a recursive algorithm, we may also reason that we need stack space to store the sequence of calls. Each stack frame does not need to hold a copy of the array. Usually, arrays are treated as references. This means that each stack frame is likely small, containing a link to the array's location and the indexes needed to keep our place in the array. This means that each recursive call will take up a constant amount of memory. The other question we need to address is "How large will the stack of frames grow during execution?" We can expect as many recursive calls as the depth of the treelike structure in the diagrams above. We know now that that depth is approximately log n. Now we have all the pieces to think about the overall space complexity of Merge Sort.

First, we need n values for the original array. Next, we will need another n value for storage in the worst case (near the end of the algorithm). Finally, we can expect the stack to take up around log n stack frames. This gives the following formula using c_1 and c_2 to account for a small number of extra variables associated with each category (indexes for the temporary arrays, start and end indexes for recursive calls, etc.): $S(n) = n + c_1*n + c_2*\log n$. This leads to $O(2*n + \log n)$, which simplifies to just O(n) in Big-O notation. The overall space complexity of Merge Sort is O(n).

Auxiliary Space and In-Place Sorting

We have now discussed the worst-case space and time complexity of Merge Sort, but an important aspect of Merge Sort still needs to be addressed. All the sorting algorithms we have discussed so far have worst-case space complexity bounds of O(n), meaning they require at most a constant multiple of the size of their input data. As we should know by now, constant values can be large and do make a difference in real-world computing. Another type of memory analysis is useful in practice. This is the issue of auxiliary memory.

It is understood that for sorting we need enough memory to store the input. This means that no sorting algorithm could require less memory than is needed for that storage. A lower bound smaller than n is not possible for sorting. The idea behind auxiliary memory analysis is to remove the implicit storage of the input data from the equation and think about how much "extra" memory is needed.

Let us try to think about auxiliary storage for Selection Sort. We said that Selection Sort only uses the memory needed for the array plus a few extra variables. Removing the array storage, we are left with the "few extra variables" part. It means that a constant number of "auxiliary" variables are needed, leading to an auxiliary space cost of O(1) or constant auxiliary memory usage. Insertion Sort falls into the same category, needing only a few index variables in addition to storage used for the array itself. Insertion Sort has an auxiliary memory cost of O(1).

An algorithm of this kind that requires only a constant amount of extra memory is called an **"in-place"** algorithm. The algorithm keeps array data as a whole within its original place in memory (even if specific values are rearranged). Historically, this was a very important feature of algorithms when memory was expensive. Both Selection Sort and Insertion Sort are in-place sorting algorithms.

Coming back to Merge Sort, we can roughly estimate the memory usage with this function:

$$S(n) = n + c_1{*}n + c_2{*}\log n.$$

Now we remove n for the storage of the array to think about the auxiliary memory. That leaves us with $c_1{*}n + c_2{*}\log n$. This means our auxiliary memory usage is bounded by $O(n + \log n)$ or just $O(n)$. This means that Merge Sort potentially needs quite a bit of extra memory, and it grows proportionally to the size of the input. This represents the major drawback of Merge Sort. On modern computers, which have sizable memory, the extra memory cost is usually worth the speed up, although the only way to know for sure is to test it on your machine.

Quick Sort

The next sorting algorithm we will consider is called Quick Sort. Quick Sort represents an interesting algorithm whose worst-case time complexity is $O(n^2)$. You may be thinking, "n^2? We already have an $O(n \log n)$ algorithm. I'll pass, thank you." Well, hold on. In practice, Quick Sort performs as well as Merge Sort for most cases, but it does much better in terms of auxiliary memory. Let us examine the Quick Sort algorithm and then discuss its complexity. This will lead to a discussion of average-case complexity.

Description of Quick Sort

The general idea of Quick Sort is to choose a pivot key value and move any array element less than the pivot to the left side of the array (for increasing or ascending order). Similarly, any value greater than the pivot should move to the right. Now on either side of the pivot, there are two smaller unsorted portions of the array. This might look something like this: **[*all numbers less than pivot*, the pivot value, *all numbers greater than pivot*]**. Now the pivot is in its correct place, and the higher and lower values have all moved closer to their final positions. The next step recursively sorts these two portions of the array in place. The process of moving values to the left and right of a pivot is called "partitioning" in this context. There are many variations on Quick Sort, and many of them focus on clever ways to choose the pivot. We will focus on a simple version to make the runtime complexity easier to understand.

Quick Sort Implementation

To implement Quick Sort, we will use a few helper functions. We have already seen the first helper function, exchange (above). Next, we will write a partition function that does the job of moving the values of the array on either side of the pivot. Here start is the first and end is the last valid index in the array. For example, end would be n – 1 (rather than n) when partitioning the whole array. This will be important as we recursively sort each subset of values with Quick Sort.

```
function partition(array, start, end)
    # make the first value the pivot
    set pivotValue to array[start]
    set smallIndex to start
    for index from start + 1 to end
        if array[index] < pivotValue
            set smallIndex to smallIndex + 1
            exchange(array, smallIndex, index)
    # exchange the pivot with the last small value
    exchange(array, start, smallIndex)
    return smallIndex
```

This partition function does the bulk of the work for the algorithm. First, the pivot is assumed to be the first value in the array. The algorithm then places any value less than the pivot on the left of the eventual position of the pivot value. This goal is accomplished using the **smallIndex** value that holds the position of the last value that was smaller than the pivot. When the loop advances to a position that holds a value smaller than the pivot, the algorithm exchanges the smaller value with the one to the right of **smallIndex** (the rightmost value considered so far that is smaller than the pivot). Finally, the algorithm exchanges the first value, the pivot, with the small value at **smallIndex** to put the pivot in its final position, and **smallIndex** is returned. The final return provides the pivot value's index for the recursive process that we will examine next.

```
function recursiveQuickSort(array, start, end)
    if start < end
        set pivotIndex to partition(array, start, end)
        recursiveQuickSort(array,start, pivotIndex - 1)
        recursiveQuickSort(array,pivotIndex + 1, end)
```

Using recursion, the remainder of the algorithm is simple to implement. We will recursively sort by first partitioning the values between start and end. By calling partition, we are guaranteed to have the pivot in its correct position in the array. Next, we recursively sort all the remaining values to the left and right of the pivot location. This completes the algorithm, but we may wish to create a nice wrapper for this function to avoid so much index passing.

```
function quickSort(array)
    set start to 0
    set end to length(array) - 1
    recursiveQuickSort(array,start, end)
```

Quick Sort Complexity

The complexity analysis of Quick Sort is interesting. We know that Quick Sort is a recursive algorithm, so we may reason that its complexity is like Merge Sort. One advantage of Quick Sort is that it is "in place." There are no copies of the data array, so there should not be any need for extra or "auxiliary" space. Remember though, we do need stack space to handle all those recursive calls and their local index variables. The critical question now is "How many recursive calls can we expect?" This question will determine our runtime complexity and reveal some interesting features of Quick Sort and Big-O analysis in general.

To better understand how this process may work, let's look at an example using the same array we used with Merge Sort.

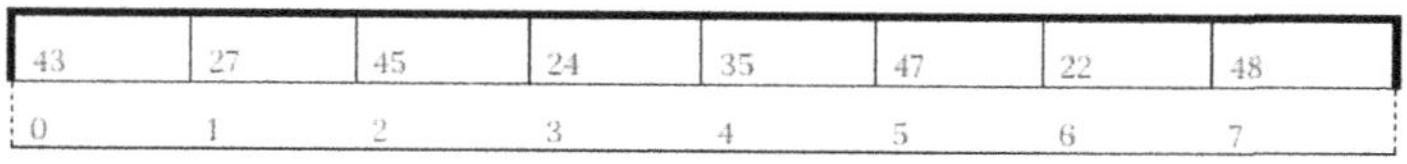

Figure 3.29

Quick Sort would begin by calling **partition** and setting the **pivotValue** to 43. The **smallIndex** would be set to 0. The loop would begin executing with **index** at 1. Since the value at **index** (27) is less than the pivot (43), the algorithm would increment **smallIndex** and exchange it with the value at **index**. In this case, nothing happens, as the indexes are the same. This is OK. Trying to optimize the small issues would substantially complicate the code. We are striving for understanding right now.

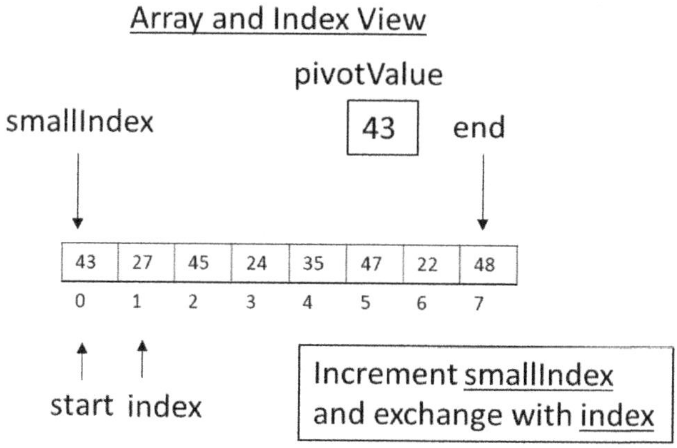

Figure 3.30

With **smallIndex** updated, execution continues through the loop examining new values. Since 45 is greater than the pivot, the loop simply continues and updates **index** only.

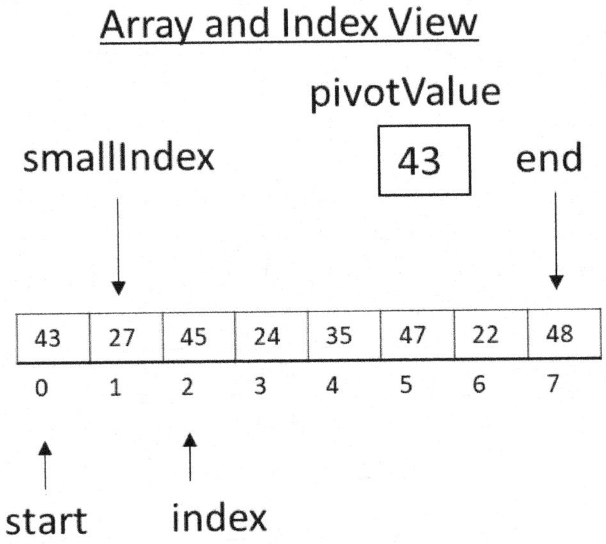

Figure 3.31

The next value, 24, is smaller than our pivot. This means the algorithm updates **smallIndex** and exchange values.

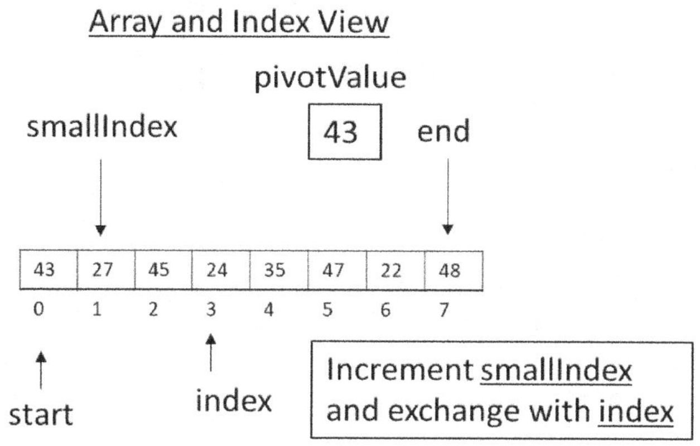

Figure 3.32

Now the array is in the following state just before the loop finishes. Next, the **index** will be incremented, and position 4 will be examined. Notice how **smallIndex** always points to the rightmost value smaller than the pivot.

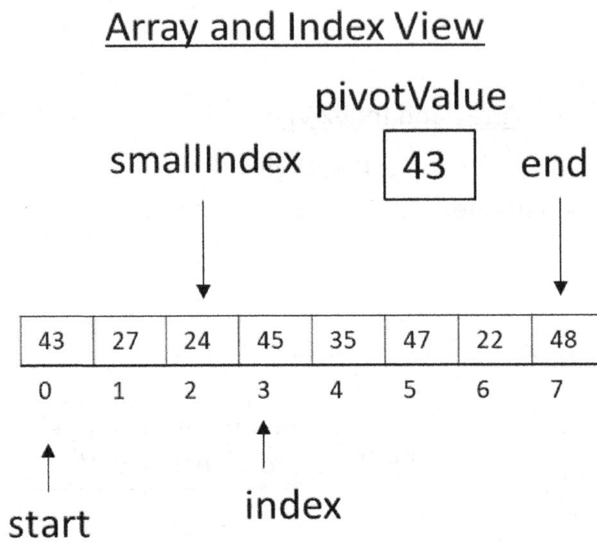

Figure 3.33

The **partition** function will continue this process until **index** reaches the end of the array. Once the loop has ended, the pivot value at the start position will be exchanged with the value at the **smallIndex** position.

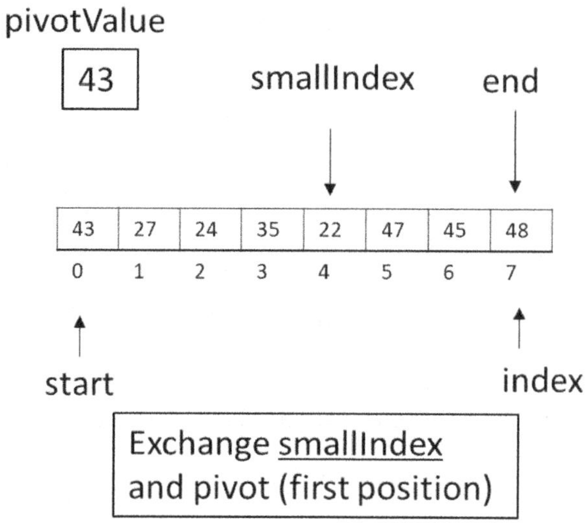

Figure 3.34

Now the **partition** function is complete, and the next step in the algorithm is to recursively Quick Sort the left and right partitions. In the figure below, we use **startL** and **endL** to mark the start and end of the left array half (and similarly for the right half):

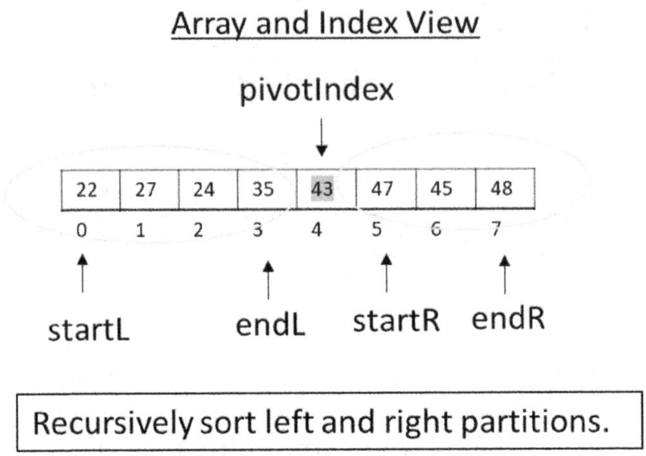

Figure 3.35

Now the process of partitioning would begin again for both sides of the array. You can envision this process growing like a tree with each new partition being broken down into two smaller parts.

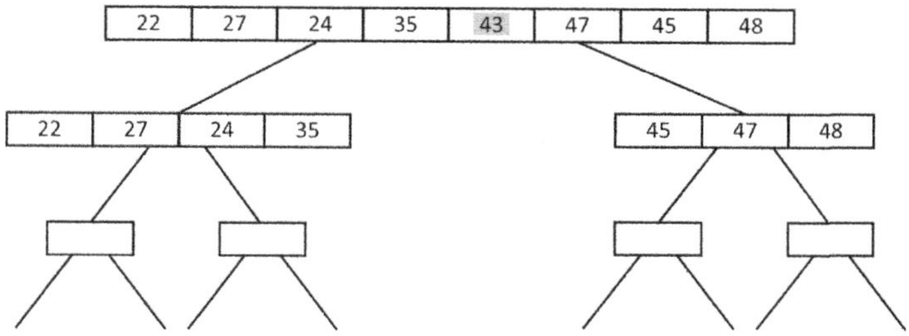

Figure 3.36

Let's assume that each recursive call to **partition** breaks the array into roughly equal-size parts. Then the array will be roughly split in half each time, and the tree will appear to be balanced. If this is the case, we can now think about the time complexity of this algorithm. The **partition** function must visit every value in the array to put it on the correct side of the pivot. This means that **partition** is O(n). Once the array is split, partition runs on two smaller arrays each of size n/2 as we assumed. For simplicity, we will ignore subtracting 1 for the pivot. This will not change the Big-O complexity. This splitting means that the second level of the tree needs to process two calls to **partition** with inputs of n/2, so 2(n/2) or n. This reasoning is identical to our analysis of Merge Sort. The time complexity is then determined by the height of this tree. In our analysis of Merge Sort, we determined that a balanced tree has a height of log n. This leads to a runtime complexity of O(n log n).

So if partitioning roughly splits the array in half every time, the time complexity of Quick Sort is O(n log n). That is a big "if" though. Even in this example, we can see that this is not always the case. We said that for simplicity we would just choose the very first value as our pivot. What happens on the very next recursive iteration of Quick Sort for our example?

For the left recursive call, 22 is chosen as the pivot, but it is the smallest value in its partition. This leads to an uneven split. When Quick Sort runs recursively on these parts, the left side is split unevenly, but the right side is split evenly in half.

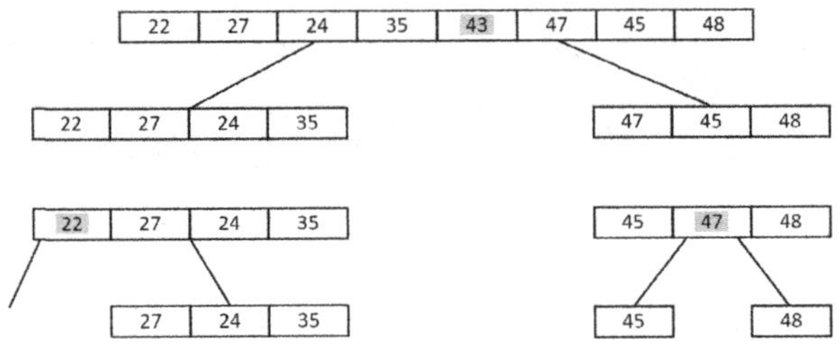

Figure 3.37

The possibility of uneven splits hints that more work might be required. This calls into question our optimistic assumption that the algorithms will split the array evenly every time. This shows that things could get worse. But how bad could it get? Let's think about the worst-case scenario. We saw that when 22 was the smallest value, there was no left side of the split. What if the values were 22, 24, 27, 35?

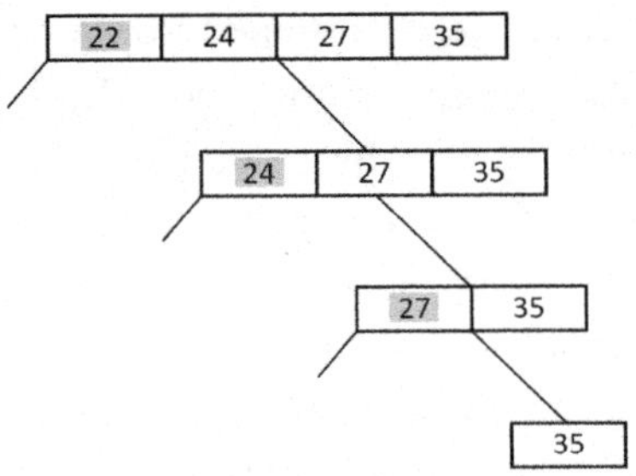

Figure 3.38

This shows an array that is already sorted. Let's now assume that we have a list that is already sorted, and every time a pivot is chosen, it chooses the smallest value. This means that the partitions that are created are an empty left subset and a right subset that contains all the remaining values minus the pivot. Recursion runs Quick Sort on the remaining n - 1 values. This process would produce a very uneven tree. First, we run **partition** on n elements, then n - 1 elements, then n - 2, and so on.

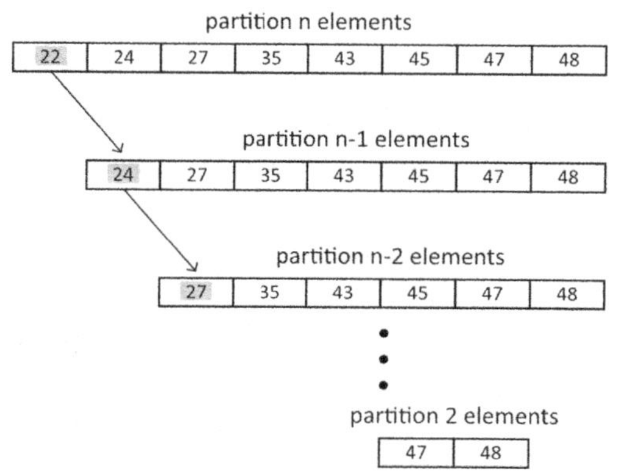

Figure 3.39

This process produces a progression that looks like our analysis of Insertion Sort but in reverse. We get n + n - 1 + n - 2 + ... 3 + 2. This leads to a complexity of $O(n^2)$. This worst-case runtime complexity is $O(n^2)$. You may be thinking that "quick" is not a great name for an algorithm with a quadratic runtime. Well, this is not the full story either. In practice, Quick Sort is very fast. It is comparable to Merge Sort in many real-world settings, and it has the advantage of being an "in-place" sorting algorithm. Let's next explore some of these ideas and try to understand why Quick Sort is a great and highly used algorithm in practice even with a worst-case complexity of $O(n^2)$.

Average-Case Time Complexity

We now know that bad choices of the pivot can lead to poor performance. Consider the example Quick Sort execution above. The first pivot of 43 was near the middle, but 22 was a bad choice in the second iteration. The choice of 47 on the right side of the second iteration was a good choice. Let us assume that the values to be sorted are randomly distributed. This means that the probability of choosing the worst pivot should be 1/n. The probability of choosing the worst pivot twice in a row would be 1/n * 1/(n - 1). This probability is shrinking rapidly.

Here is another way to think about it. The idea of repeatedly choosing a bad pivot by chance is the same as encountering an already sorted array by chance. So the already sorted order is one ordering out of all possible orderings. How many possible orderings exist? Well, in our example we have 8 values. We can choose any of the 8 as the first value. Once the first value is chosen, we can choose any of the remaining 7 values for the second value. This means that for the first two values, we already have 8*7 choices. Continuing this process, we get 8*7*6*5*4*3*2*1. There is a special mathematical function for this called **factorial**. We represent factorial with an exclamation point (!). So we say that there are 8! or 8 factorial possible orderings for our 8 values. That is a total of 40,320 possible orderings with just 8 values! That means that the probability of encountering by chance an already sorted list of 8 values is 1/8! or 1/40,320, which is 0.00002480158. Now imagine the perfectly reasonable task of sorting 100 values. The value of 100! is greater than the estimated number of atoms in the universe! This makes the probability of paying the high cost of $O(n^2)$ extremely unlikely for even relatively small arrays.

To think about the average-case complexity, we need to consider the complexity across all cases. We could reason that making the absolute best choice for a pivot is just as unlikely as making the absolute worst choice. This means that the vast majority of cases will be somewhere in the middle. Researchers have studied Quick Sort and determined that the average complexity is O(n log n). We won't try to formally prove the average case, but we will provide some intuition for why this might be. Sequences of bad choices for a pivot are unlikely. When a pivot is chosen that partitions the array unevenly, one part is smaller than the other. The smaller subset will then terminate more quickly during recursion. The larger part has another chance to choose a decent pivot, moving closer to the case of a balanced partition.

As we mentioned, the choice of pivot can further improve the performance of Quick Sort by working harder to avoid choosing a bad pivot value. Some example extensions are to choose the pivot randomly or to select 3 values from the list and choose the median. These can offer some improvement over choosing the first element as the pivot. Other variations switch to Insertion Sort once the size of the partitions becomes sufficiently small, taking advantage of the fact that small data sets may often be almost sorted, and small partitions can take advantage of CPU cache efficiency. These modifications can help improve the practical measured runtime but do not change the overall Big-O complexity.

Quick Sort Space Complexity

We should also discuss the space complexity for Quick Sort. Quick Sort is an "in-place" sorting algorithm. So we do not require any extra copies of the data. The tricky part about considering the space complexity of Quick Sort is recognizing that it is a recursive algorithm and therefore requires stack space. As with the worst-case time complexity, it is possible that recursive calls to Quick Sort will require stack space proportional to n. This leads to O(n) elements stored in the array and O(n) extra data stored in the stack frames. In the worst case, we have O(n + n) total space, which is just O(n). Of the total space, we need O(n) auxiliary space for stack data during the recursive execution of the algorithm. This scenario is unlikely though. The average case leads to O(n + log n) space if our recursion only reaches a depth of log n rather than n. This is still just O(n) space complexity, but the auxiliary space is now only O(log n).

Exercises

1.
Create a set of tools to generate random arrays of values for different sizes in your language of choice. These can be used to test your sorting algorithms. Some useful functions and capabilities are provided below.

   a. Include functions to generate random arrays of a given size up to 10,000.
   b. Include functions to print the first 5 numbers or the entire array.
   c. Provide parameters to control the range of values.
   d. Provide functions to generate already sorted or reverse sorted arrays.
   e. Explore your programming language's time functionality to be able to measure sorting performance in terms of the time taken to complete

the sort.

f. Consider creating functions to repeatedly run an algorithm and record the average sorting time.

2.

Implement Selection Sort and Insertion Sort in your language of choice. Using randomly generated array data, try to find the number of values where Insertion Sort begins to improve on Selection Sort. Remember to repeat the sorting several times to calculate an average time.

3.

Implement Merge Sort and repeat the analysis for Merge Sort and Insertion Sort. For what n does Merge Sort begin to substantially improve on Insertion Sort? Or does it seem to improve at all?

4.

Implement Quick Sort in your language of choice. Next, determine the time for sorting 100 values that are already sorted using Quick Sort (complete the 1.d exercise). Next, randomly generate 1,000 arrays of size 100 and sort them while calculating the runtime. Are any as slow as Quick Sort on the sorted array? If so, how many? If not, what is the closest time?

References

JaJa, Joseph. "A Perspective on Quicksort." *Computing in Science & Engineering* 2, no. 1 (2000): 43–49.

4. Search

Learning Objectives

After reading this chapter you will...

- understand how to find specific data within an array by searching.
- be able to implement a search algorithm that runs in O(n) time.
- be able to implement Binary Search for arrays that will run in O(log n) time.

Introduction

The problem of finding something is an important task. Many of us will spend countless hours in our lives looking for our keys or phone or trying to find the best tomatoes at the grocery store. In computer science, finding a specific record in a database can be an important task. Another common use for searching is to check if a value already exists in a collection. This function is important for implementing "sets" or collections of unique elements.

With the search problem, we start to think a little more about our data structures and what a solution means in the context of the data structure used. Suppose, for example, that we have an array of the following values using 0-based indexing.

43	27	45	24	35	47	22	48
0	1	2	3	4	5	6	7

Figure 4.1

What does it mean to find the value 22? Should we return True if 22 is in the array? Should we return the index 6 instead? What should we do if 22 is not found? These questions will depend on how the search function is used in its broader context.

Let us describe a search problem in more detail. First, we are looking for a specific value or a record identified by a specific code in our data set. We will call the value for which we are searching the "key." The key is a data value used to find a match in the data structure. For a simple array, the key is just the value itself. For example, 22 could be the key for which we are looking. Further, we may specify that our algorithms should return True if the key is found in our data structure and **False** if we fail to find the key. With our previous array, a call to **search(array, 22)** should return True, but a call to **search(array, 12)** should return False.

We now have an idea of what our search function should do, but we do not yet have an idea of how it should do it. Can you think of a way to implement search? Take a few minutes to think about it.

I am sure most computer science students would come to the same idea. Examine all the values of the array one by one, which is also known as iterating. If one of the values matches the key, return True. If we get to the end without finding the key, return False. This is a simple idea that will definitely work. This is the strategy behind **Linear Search**, which we will examine in the next section.

Linear Search

Linear Search may be the simplest searching algorithm. It uses an approach similar to the way a human might look for something in a systematic way—that is, by examining everything one by one. If your mother puts your clothes away and you are trying to find your favorite shirt, you might try every drawer in your room until it is found. Linear Search is an exhaustive search that will eventually examine every value in an array one by one. You can remember the name *linear* by thinking of it as going one by one in a *line* through all the values. Linear is also a clue that the runtime is O(n) because in the worst case, we must examine all n items in the array.

Let us examine one implementation of Linear Search.

```
function linearSearch(array, key)
    set end to length(array)
    for index from 0 to end - 1
        if array[index] equals key
            return True
    # After the loop, return False, value not found
    return False
```

Linear Search Complexity

As always, we will be interested in assessing the time and space scaling behavior of our algorithm. This means we want to know how its resource demand grows with larger inputs.

For space complexity, Linear Search needs storage for the array and a few other variables (the index of our for-loop, for example). This leads to a bound of O(n) space complexity.

For time complexity, we want to think about the best-case and worst-case scenarios. Suppose we go to search for the key, 22, and as luck would have it, 22 is the first value in the array! This leads to only a small number of operations and only 1 comparison operation. As you may have guessed, the best-case scenario leads to an O(1) or constant number of operations. In algorithm analysis (as in stock market investing), luck is not a strategy. We still need to consider the worst-case behavior of the algorithm, as this characteristic makes a better tool for evaluating one algorithm against another. In general, we cannot choose the problems we encounter, and our methods should be robust against all types of problems that are thrown at us. The worst case for Linear Search would be a problem where our key is found at the end of the array or isn't found at all. For inputs with this feature, our time complexity bound is O(n).

Linear Search with Objects

Suppose we designed our Linear Search function to return the actual value of the key. For the array [43, 27, 45, 24, 35, 47, 22, 48], **search(array, 22)** would return 22. An important design consideration is this: What should it return if the value is not found in the array? Some approaches would return −1, but this would limit our search values to positive integers. What is needed is some type of sentinel value. This is another special value, unlike the ones we are storing. Another approach could be to throw an exception if the value is not found. There are many ways to address this problem, and this issue is an important one when storing more complex data types than just integers.

Suppose we are working on a database of contact information for a student club. We would design a class or data type specification for the student records that we need to store in our data structures. Our Student class might look something like the code below:

```
class Student
    Integer member_id
    String name
    String email
```

Now think about storing an array of Student object instances in memory. The diagram below is one way to visualize this data structure:

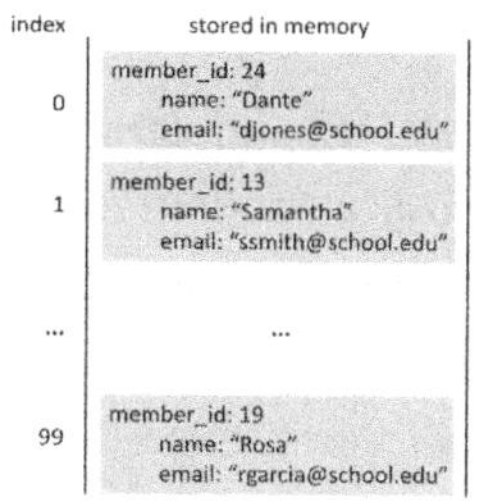

Figure 4.2

Now suppose we want to search through our database for the student whose **member_id** is 22. If our student is in the array, we could just return the Student object. If there is no student with 22 as their **member_id**, we run into the issues we mentioned above. All these issues create some difficulties in designing the interface of our search algorithm. A simple solution that could sidestep the problem would be to return either the index of the value or -1 to indicate that the value was not found. For many programming languages, -1 is an invalid array index.

Let us try our implementation of Linear Search one more time using the indexed approach and assume that our array holds a set of Student objects.

```
function linearSearch(array, key)
    set end to length(array)
    for index from 0 to end - 1
        if array[index].member_id equals key
            return index
    # After the loop, no student was found
    # return -1 as an invalid index
    return -1
```

An example use of this implementation is given below. The programmer could access a student object safely (only if it was found) using the array index after the array has been searched.

```
set index to linearSearch(database, 22)
if index > -1
    update(database[index])
else
    print "Student not found"
```

A slight variation on this idea comes from a slightly altered database. Rather than storing all our student records in continuous blocks of memory, we may have an array of references to our records. This would lead to a structure like that depicted in the following image:

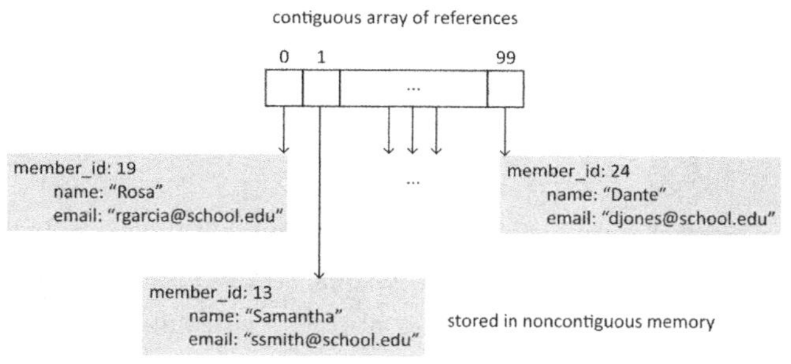

Figure 4.3

For this style of storage, our array holds references to instances rather than having the instances of objects stored in the array. Holding references rather than objects comes with some advantages in flexibility, but working with references puts more responsibility for memory management in the hands of the programmer. For the present search problem, working with references gives a nice solution to the "not found" problem. Specifically, we can return a null reference when our search fails to find an object with the matching search key.

Suppose now that our database of students is an array of references to objects. Our implementation would look like this:

```
function linearSearch(array, key)
    set end to length(array)
    for index from 0 to end - 1
        if array[index].member_id equals key
            return array[index]
    # After the loop, no student was found
    # return null as an invalid reference
    return null
```

Using this implementation as before may look something like this:

```
set reference to linearSearch(database, 22)
if reference is not null
    update(reference)
else
    print "Student not found"
```

These examples will help you appreciate how simple design questions can lead to difficult issues when implementing your algorithm. Even without thinking about performance (in terms of Big-O complexity), design issues can impact the usability and usefulness of real-world software systems. Answering these design questions will inevitably impose constraints on how your algorithm can and will be used to solve problems in real-world contexts. It is important to carefully consider these questions and to understand how to think about answering them.

Binary Search

We have seen a method for searching for a particular item in an array that runs in O(n) time. Now we examine a classic algorithm to improve on this search time. The **Binary Search** algorithm improves on the runtime of Linear Search, but it requires one important stipulation. For Binary Search to work, the array items must be in a sorted order. This is an important requirement that is not cost-free. Remember from the previous chapter that the most efficient general-purpose sorting algorithms run in O(n log n) time. So you may ask, "Is Binary Search worth the trouble?" The answer is yes! Well, it depends, but generally speaking, yes! We will return to the analysis of Binary Search after we have described the algorithm.

The logic of Binary Search is related to the strategy of playing a number-guessing game. You may have played a version of this game as a kid. The first player chooses a number between 1 and 100, and the second player tries to guess the number. The guesser guesses a number, and the chooser reports one of the following three scenarios:

1. The guesser guessed the chooser's number and wins the game.
2. The chooser's number is higher than the guess, and the chooser replies, "My number is higher."
3. The chooser's number is lower than the guess, and the chooser replies, "My number is lower."

An example dialogue for this game might go like this:
Chooser: [chooses 37 in secret] "I have my number."
Guesser: "Is your number 78?"
Chooser: "My number is lower."
Guesser: "Is your number 30?"
Chooser: "My number is higher."
Guesser: "Is your number 47?"
Chooser: "My number is lower."
Guesser: "Is your number 35?"
Chooser: "My number is higher."
Guesser: "Is your number 40?"
Chooser: "My number is lower."
Guesser: "Is your number 38?"
Chooser: "My number is lower."
Guesser: "Is your number 37?"
Chooser: "You guessed my number, 37!"

With each guess, the guesser narrows down the possible range for the chooser's number. In this case, it took 7 guesses, but if the guesser is truly guessing at random, it could take much longer. Where does Binary Search come in? Well, take a few moments to think about a better strategy for finding the right number. When the guesser guesses 78 and the chooser responds with "lower," all values from 78 to 100 can be eliminated as possibilities. What strategy would maximize the number that we eliminate each time? Maybe you have thought of the strategy by now.

The optimal strategy would be to start with 50, which eliminates half of the numbers with one guess. If the chooser responds "lower," the next guess should be 25, which again halves the number of possible guesses. This process continues to split the remaining values in half each time. This is the principle behind Binary Search, and the "binary" name refers to the binary split of the candidate values. This strategy works because the numbers from 1 to 100 have a natural order.

A precondition for Binary Search is that the elements of the array are sorted. The sorting allows each comparison in the array to be oriented, and it indicates in which direction to continue the search. Each check adds some new information for our algorithm and allows the calculation to proceed efficiently.

We will present an illustration below of an example execution of the algorithm. Suppose we are searching for the key 27 in the sorted array below:

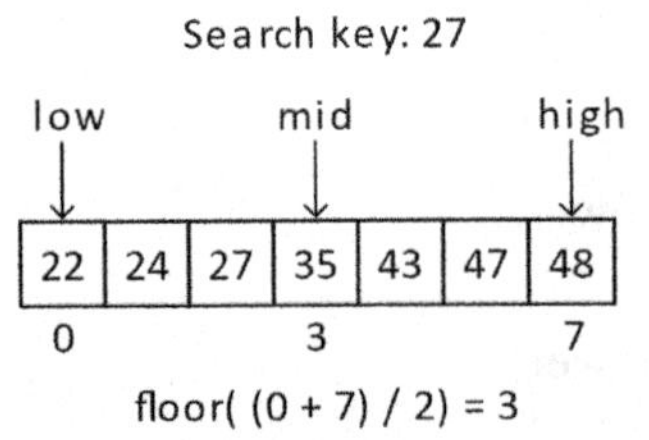

Figure 4.4

We will keep track of three index variables to track the low and high ends of the range as well as a "mid" or middle variable that will track the middle value in the range. The mid variable will be the one currently considered for the key. In this case, 35 is too high, so we will update the high end of the range. The high variable will be set to mid – 1, and we will recalculate the mid.

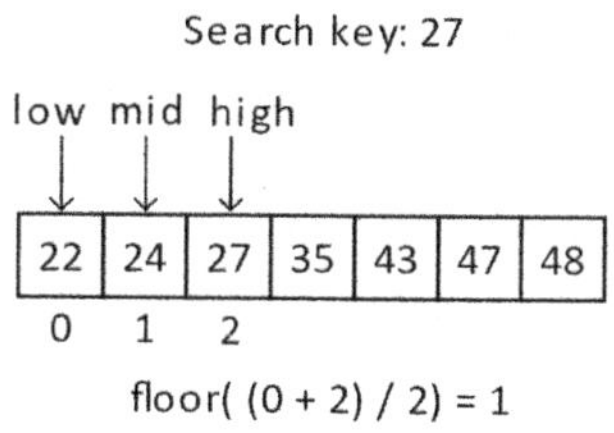

Figure 4.5

At this point in the execution, mid at 1 means we are considering the value 24. As 27 is greater than 24, we will now update the low variable to mid + 1 or 2.

Next, we have the case where low equals high, and that means either we have found the key or the key does not exist. We see that 27 is in the array, and we would report that it is found. The image below gives this scenario:

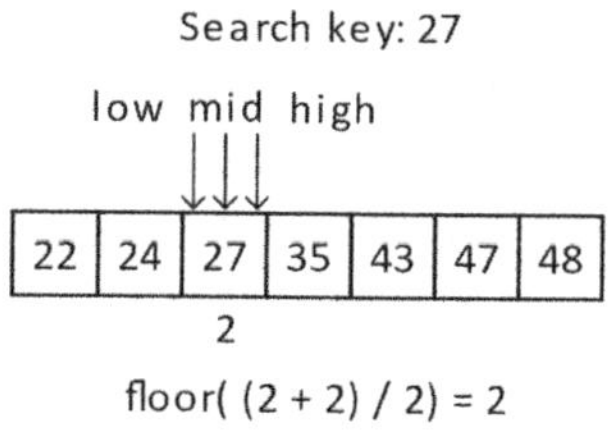

Figure 4.6

The game description and array example should give you an idea of how Binary Search can efficiently find keys in a sorted data structure. Let us examine an implementation of this algorithm. For our design, we will return the index of the value if it is found or −1 as an invalid index to indicate the key was not found. We will consider an array of integer keys, but it will work equally well with objects assuming they are sorted by their relevant keys.

```
1  # precondition: the array is sorted
2  function binarySearch(array, key)
3      set low to 0
4      set high to length(array) − 1
5      while low <= high
6          set mid to floor((low + high) / 2)
7          if key < array[mid]
8              # key is lower, move high down
9              set high to mid − 1
10         else if key > array[mid]
11             # key is higher, move low up
12             set low to mid + 1
13         else
14             # key is at mid position
15             return mid
16     # key not found
17     return −1
```

This is a subtle and powerful algorithm. It may take some thinking to understand. Think about when the algorithm would reach line 15. This means that the value at array[mid] is neither higher nor lower than the key. If it is not higher or lower, it must be the key! We return the index of the key in this case.

The case of the key missing from the array is also subtle. How could the algorithm reach line 17? To reach line 17, low must be a value greater than high. How could this happen? Think back to our example above when low, high, and mid were all pointing to index 2 and we were searching for the key 27. Suppose instead of 27 at position 2, the array had 25 at position 2, which still preserved the sorted order (22, 24, 25, 35, 43, 45, 47, 48). The algorithm would check "Is 27 less than 25?" at line 7. No, this is false. Next, it would check "Is 27 greater than 7?" at line 10. This is true, so low would be updated to mid + 1, or 3 in this case, and the loop would begin again. Only now, low is 3 and high is 2, and the loop condition fails and the return −1 at line 17 is reached.

Binary Search Complexity

Now we will assess the complexity of Binary Search. The space complexity of Binary Search is O(1) or constant space. From another perspective, one might consider this auxiliary space and say that O(n) space is needed to hold the data. From our perspective, we will assume that the database is needed already for other purposes and not consider its O(n) space cost a requirement of Binary Search. We will only consider the space demand for the algorithm to be the few extra variables that serve as the array indexes. Specifically, our algorithm only uses the low, high, and mid indexes. We could also factor in a reference for the array's position in memory and a copy of the key value. Even with these extra variables consuming space, only a constant amount of extra memory is needed, leaving the space complexity of the array-based Binary Search at O(1).

The time complexity of Binary Search requires a bit of explanation, but the logic behind the proof is similar to arguments we have seen before (see "Powers of 2 in O(log n) Time" in chapter 2 and "Merge Sort Complexity" in chapter 3). First, consider the best-case scenario. The best case would be if the key item is found at the mid position on the very first check. In our example above, this would occur if our key was 35, which is in position 3 (mid = floor((0 + 7) / 2) = 3). In the best case, the time complexity of Binary Search is O(1). This matches the best case for Linear Search. In the worst case, Binary Search must continue to update the range of possible locations for the key. This update process essentially eliminates half of the range each time our loop runs. This means that determining the Big-O complexity for Binary Search depends on determining how many times we can halve the range before reaching a single element. Here we have repeated division by two, which we should now know leads to O(log n). We will present this a little more formally below. Letting T(n) be the time cost in the number of operations for Binary Search on an array of N elements,

$$T(n)=c + T(n/2).$$

Here c is a constant number of operations (making a comparison, updating a value, and so on; it may be different on different computer architectures).

We can expand this like so:

$$\begin{aligned} T(n) &= c + (c + T(n/4)) \\ &= 2c + T(n/2^2) \\ &= 2c + (c + T(n/2^3)) \\ &= 3c + T(n/2^3). \end{aligned}$$

This leads us to the following formula:

$$T(n)=k*c + T(n/2^k).$$

Ultimately, repeatedly reducing the range of valid choices will lead to a single element that must be compared with the key. So we want to find the k that makes $n/2^k$ equal to 1. This value is $\log_2 n$, which we will abbreviate to just log n.

Substituting back into the equation gives the following:

$$T(n)=(\log n)*c + T(n/2^{\log n})$$
$$=(\log n)*c + T(n/n)$$
$$=(\log n)*c + c.$$

We are left with a constant multiple of log n for a worst-case time complexity of O(log n).

A time complexity of O(log n) is considered extremely fast in most contexts and is an excellent scaling bound for an algorithm. Consider a Linear Search with 1,000 items. That algorithm may have to make nearly 1,000 comparison checks to determine if the key is found. A Binary Search for a sorted array of 1,000 items needs to make only about 10 checks. For an array of 1,000,000 elements, the Linear algorithm may make nearly 1 million checks, while the Binary Search checks only about 20 in the worst case! That is an excellent improvement (1 million >> 20).

Binary Search Complexity in Context

We are all ready to celebrate and embrace the amazing properties of Binary Search with its O(log n) search time complexity, but there is a catch. As we mentioned, the array must be sorted, and typically we cannot do much better than O(n log n) for sorting (without some extra information). Would that mean that, in reality, Binary Search is O(n log n + log n) leading to O(n log n)? In a sense, yes. If we had to start from an unsorted array, we would need to first sort it. This would give us a sorting cost of O(n log n). Then any subsequent search on the data would only cost O(log n). This would make the total cost of Binary Search bounded by its most expensive operation, the sorting part. Oh no, Binary Search is actually O(n log n)—all is lost! Well, let us use our analysis skills to try to determine why and when Binary Search would be more useful than Linear Search.

The important realization is that sorting is a one-time cost. Once the array is sorted, all subsequent searches can be done in O(log n). Let us think about how this compares to Linear Search, which always has a cost of O(n) regardless of the number of the number of times the array is searched. Another name for the act of searching is called a **query**. A query is a question, and we are asking the data structure the question "Do you have the information we need?" Suppose that the variable Q is the number of queries that are made of the data structure.

Querying our array using Linear Search Q times would give the following time cost with c being a constant associated with O(n):

$$T_{LS}(n, Q)=Q * c * n.$$

Querying our array using Binary Search Q times would give the following cost:

$$T_{BS}(n. Q)=c*(n * \log n) + Q * c * (\log n).$$

Now suppose that Q was close to the size of n. We could rewrite these like this:

$$T'_{LS}(n) = n * c * n$$
$$= c * n^2.$$

This leads to a time complexity of $O(n^2)$ for searching with approximately n different queries.

For Binary Search, we have the following adjusted formula:

$$T'_{BS}(n) = c * (n * \log n) + n * c * (\log n)$$
$$= 2 * c * (n \log n).$$

This leads to a time complexity of $O(n \log n)$ for searching with approximately n different queries.

This means that if you plan on searching the data structure n or more times, Binary Search is the clear winner in terms of scalability.

As a final note, you should always try to run empirical tests on your workloads and hardware to draw conclusions about performance. Processor implementations on modern computers can further complicate these questions. For example, the CPU's branch prediction and cache behavior may make *Linear Search* on a *sorted* list faster than some clever algorithmic search implementation in terms of actual runtimes.

Exercises

1. Implement a Linear Search in your language of choice. Use the following plan to test your implementation on an array of 100 randomly generated values (in random order). Randomly generate 100 values, and use Linear Search to find the value 42. Have your search print the number of unsuccessful checks before finding the value 42 (or reporting not found).
2. Take the search function from exercise 1, and modify it to count and return the number of checks Linear Search takes to find the value 42 in a random array. Write a loop to repeat this experiment 100 times, and average the number of checks it takes to find a specific value. What is that number close to? How does it change if you increase the number of tests from 100 to 1,000?
3. The reasoning used to determine the time complexity of Binary Search closely resembles similar arguments from chapter 2 on recursion. Implement Binary Search as a recursive algorithm by adding extra parameters for the high and low variables. Make sure your function is tail-recursive to facilitate tail-call optimization.
4. With your implementations of Linear and Binary Search, write some tests to generate a number of random queries. Calculate the total time to conduct n/2 queries on a randomly generated dataset. Be sure to include the

sorting time for your Binary Search database before calculating the total time for all queries. Compare your result to the Linear Search total query time. Next, repeat this process for n, 2*n, and 4*n queries. At what number of queries does Sorting + Binary Search start to show an advantage over Linear Search?

References

Due to the age and simplicity of these algorithms, many of the published works in the early days of computing refer to them as being "well known." Donald Knuth gives some early references to their origin in volume 3 of *The Art of Computer Programming.*

Knuth, Donald E. *The Art of Computer Programming.* Pearson Education, 1997.

5. Linked Lists

Learning Objectives

After reading this chapter you will…

- begin to understand how differences in data structures result in trade-offs and help when choosing which to apply in a real-world scenario.
- begin to use links or references to build more complex data structures.
- grasp the power and limitations of common arrays.

Introduction

You have a case of cola you wish to add to your refrigerator. Your initial approach is to add all colas to the refrigerator while still in the box. That way, when you want to retrieve a drink, they are all in the same place, making them easier to find. You will also know when you are running low because the box will be nearly empty. Storing them while still in the box clearly has some benefits. It does come with one major issue though. Consider a refrigerator filled with groceries. You may not have an empty spot large enough to accommodate the entire case of cola. However, if you open the case and store each can individually, you can probably find some spot for each of the 12 cans. You have now found a way to keep all cans cold, but locating those cans is now more difficult than it was before. You are now faced with a trade-off: Would you rather have all cans cold at the cost of slower retrieval times or all cans warm on the counter with faster retrieval times? This leads to judgment calls, like deciding between how much we value a cold cola and how quickly we need to retrieve one.

Our case of cola is like a data structure, and storing all cans in the box is analogous to an array. Just like the analogy, let us start by listing some of the desirable characteristics of arrays.

- We know exactly how many elements reside in them, both now and in the future. We know this because (in most languages) we are required to specify the length explicitly. Also, most implementations of arrays do not allow us to simply resize as needed. As we will see soon, this can be both a beneficial feature and a constraint.
- They are fast. Arrays are indexable data structures with lookups in constant time. In other words, if you have an array with 1,000 elements, getting the value at index 900 does not mean that you must first look at the first 899 elements. We can do this because array implementations rely on storage within contiguous blocks of memory. Those contiguous blocks have predictable and simple addressing schemes. If you are standing at the beginning of an array (say at address 0X43B), you can simply multiply 900 by the size of the element type stored in the array and look up the memory location that is that distance from the starting point. This can be done in constant time, O(1).

These desirable characteristics are also constraints if you look at them from a different perspective.

- Having an explicit length configured before you use the array does mean that we know the length without having to inspect the data structure, but it also means that we cannot add any new elements once we reach the capacity of the array. For plenty of applications, we may not know the proper size before we begin processing.

- Arrays are fast because they are stored in contiguous blocks of memory. However, for really large sets of data, it may be expensive (regarding time) or impossible (regarding space) to find a sufficiently large contiguous block of memory. In these cases, an array may perform poorly or not at all.

It is clear that, under certain circumstances, arrays may not serve all our needs. We now have a motivation for new types of data structures, which bring with them new trade-offs. The first of these new data structures that we will consider is the linked list.

Structure of Linked Lists

Linked lists are the first of a series of reference-based data structures that we will study. The main difference between arrays and linked lists is how we define the structure of the data. With arrays, we assumed that all storage was contiguous. To locate the value at index 5, we simply take the address of the beginning of the array and add 5 times the size of the data type we are storing. That gives us the address of the data we wish to retrieve. Of course, most modern languages give us simpler indexing operators to accomplish the task, but the description above is essentially what happens at a lower level.

Linked lists do not use contiguous memory, which implies that some other means of specifying a sequence must be used. Linked lists (along with trees and graphs in later chapters) all use the concept of a node, which references other nodes. By leveraging these references and adding a value (sometimes called a payload), we can create a data structure that is distinctive from arrays and has different trade-offs.

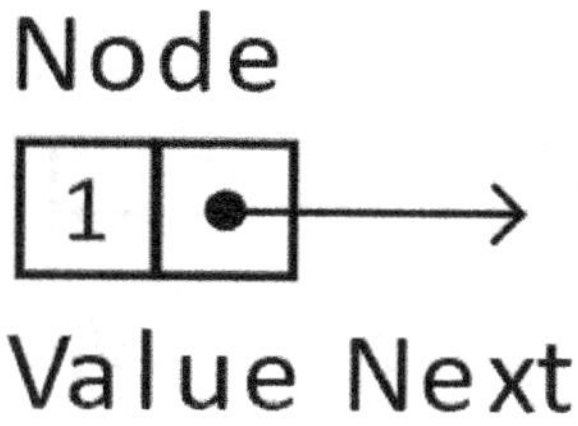

Figure 5.1

When working with the linked list, the next element in the structure, starting from a given element, is determined by following the reference to the next node. In the example below, node a references (or points to) node b. To determine the elements in the structure, you can inspect the payloads as you follow the references. We follow these references until we find a null reference (or a reference that points to nothing). In this case, we have a linked list of length 2, which has the value 1 followed by the value 12.

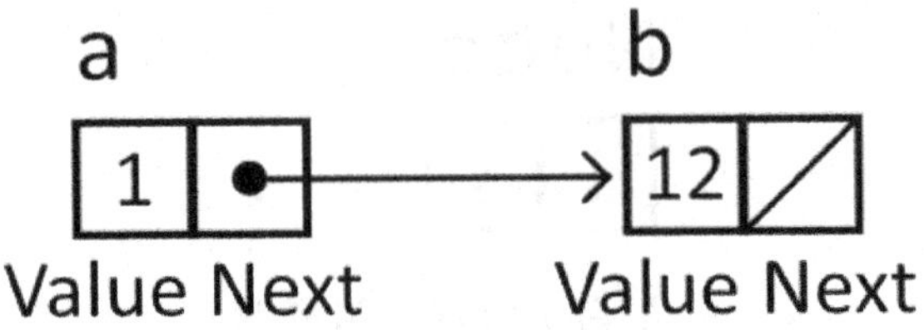

Figure 5.2

From a practical standpoint, implementations require that the chosen language have some sort of feature that allows for grouping the value with the reference. Most languages will accomplish this task with either structs or objects. For pseudocode examples, we will assume the following definition of a node. The payload is of type integer because it is convenient for the remainder of the chapter, but the data stored in the node can be of any type that is useful given some real-world circumstances.

```
class Node
    Integer Value
    Node Next
```

Let us consider the following explicitly defined list with powers of 3 as the values. The choice of values is intended to reinforce the sequential nature of the data structure and could have easily been any other well-known sequence. The critical step is how the **Next** reference is assigned to subsequent nodes in lines 9, 10, and 11. Later, we will define a procedure for inserting values at an arbitrary position, but for now, we will use **a** as our root. The resulting linked list is depicted in figure 5.3.

```
1  set a to a new instance of Node
2  set a.Value to 1
3  set b to a new instance of Node
4  set b.Value to 3
5  set c to a new instance of Node
6  set c.Value to 9
7  set d to a new instance of Node
8  set d.Value to 27
9  set a.Next = b
10 set b.Next = c
11 set c.Next = d
```

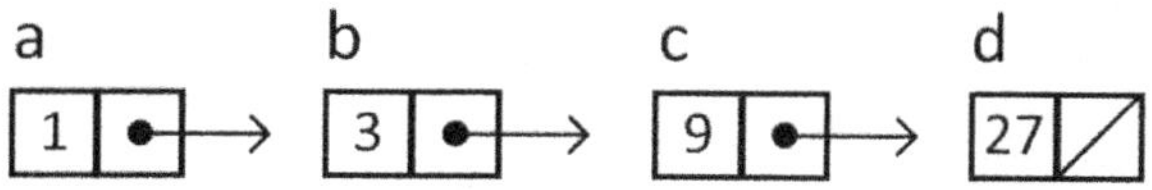

Figure 5.3

Operations on Linked Lists

Lookup (List Traversal)

Continuing the comparison with arrays, our first task will be to look up the value at an arbitrary position relative to a particular node in a linked list. You may consider a position in a linked list as the analogue of an array's index. It is the ordinal number for a specific value or node within the data structure. For the sake of consistency, we will use 0-based positions.

```
1 function lookup(rootNode, position)
2     set n to rootNode
3     for i = 0 to position
4         n = n.Next
5     return n.Value
```

We address lookup first because it most clearly illustrates the means by which we traverse the linked list. On line 2, we start with some node, then lines 3 and 4 step forward to the next node an appropriate number of times. Whenever we wish to insert, delete, or look up a value or simply visit each node, we must either iteratively or recursively follow the **Next** reference for the desired number of sequential nodes.

Now that we have a means of looking up a value at an arbitrary position, we must consider how it performs. We start again by considering arbitrary index lookups in arrays. Recall that a lookup in an array is actually a single dereference of a memory address. Because dereferencing on an array is not dependent on the size of the array, the runtime for array lookups is O(1). However, to look up an arbitrary element in a linked list (say, the nth element), we must dereference n – 1 different addresses as we follow the **Next** reference. We now have a number of dereferences dependent on the length of the linked list. Specifically, our cost function of the worst-case scenario will be f(n) = n – 1, where n is the length, which is clearly O(n). Dereferencing within some loop or with recursion is a featured pattern in nearly every linked list algorithm. Therefore, in most cases, we will expect these algorithms to run in O(n) time.

Length (and Additional Means of Traversal)

While it is the case that most list traversals are implemented with for-loops, there are occasions where other styles of traversal are more appropriate. For example, for-loops are ill-suited for scenarios where we do not know exactly how many times we must loop. As a result, while-loops are often used whenever all nodes must be visited. Consider the function below, which returns an integer representing the number of elements in the list starting at **rootNode**:

```
function len(rootNode)
    set n to rootNode
    set c to 0
    while n is not null
        set n to n.Next
        set c to c + 1
    return c
```

Also worth noting is that, due to the self-referencing definition of the **Node** class, many list procedures are reasonably implemented using recursion. If you consider a given root node, the length of a linked list starting at that node will be the sum of 1 and the length of the list starting at the **Next** reference (the recursive case). The length of a list starting with a null node is 0 (the base case).

```
function len(node)
    if node is null
        return 0
    else
        return 1 + len(node.Next)
```

As we explore more algorithms in this book, we will discover that often recursive solutions drastically reduce the complexity of our implementation. However, we should pay close attention here. Because we dereference the **Next** node for every node visited, our solution still runs in O(n) time.

Insert

To create a general-purpose data structure, our next operation will be to insert new values at arbitrary positions within the linked list. For the sake of simplicity, this function assumes that the position is valid for the current length of the linked list.

```
function insertAtPosition(rootNode, position, value)
    # create a node to hold the new value
    set newNode to an instance of Node
    set newNode.Value to value

    if position equals 0
        set newNode.Next to rootNode
        return newNode
    else
        # follow Next until in position to insert
        set n to rootNode
        set i to 1
        while i < position
            set n to n.Next
            set i to i + 1
        set newNode.Next to n.Next
        set n.Next to newNode
        return rootNode
```

When reading and trying to comprehend this algorithm, we should pay close attention to three key things:

- We again see a linear traversal through the list to a given position by traversing the **Next** reference. As usual, it does not matter whether this is achieved with a for-loop, a while-loop, or recursion, the result is still the same. We maneuver through a list by starting at some point and following the references to other nodes. This is an incredibly important concept, as it lays the foundation for more interesting and useful data structures later in this book.

- When implementing algorithms, edge cases sometimes require specific attention. This function is responsible for inserting the value at a desired position, regardless of whether that position is 0, 2, or 200. Inserting a value into the middle of a linked list means that we must set references for the prior's **Next** as well as **newNode.Next**. Inserting at position 0 is fundamentally different in that there is no prior node.
- More so than other statements, lines 14 and 15 may feel interchangeable. They are not. Much like the classic exchange exercise in many programming textbooks, executing these statements in the reverse order will lead to different behavior. It is a worthwhile exercise to consider what the outcome would be if they were switched.

We can visually trace the following example of an insertion:

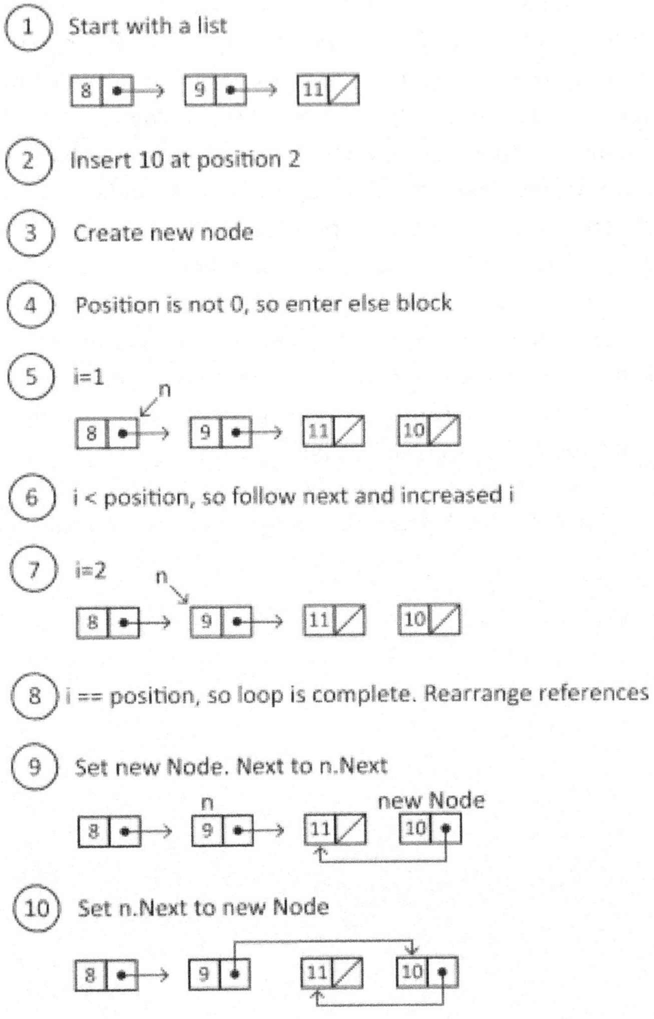
1 Start with a list
8
9
11
2 Insert 10 at position 2
3 Create new node
4 Position is not 0, so enter else block
5 i=1
n
8
9
11
10
6 i < position, so follow next and increased i
7 i=2
n
8
9
11
10
8 i == position, so loop is complete. Rearrange references
9 Set new Node. Next to n.Next
n
new Node
8
9
11
10
10 Set n.Next to new Node
8
9
11
10

Figure 5.4

Next, we come to the runtime analysis of this function. Due to the linear traversal, we consider the algorithm itself to be of O(n) regardless of whether we are inserting at position 2 or 200. However, what if we want to insert at position 0? In this case, the number of operations required is not dependent on the length of the linked list, and therefore this specific case runs in O(1). When studying algorithms, we typically categorize using the worst-case scenario but may specify edge-case runtimes when appropriate. In other words, if we only ever care about inserting at the front of a linked list, we may consider this special case of insert to be an O(1) operation.

Remove

Now that we have seen how to traverse the list via the Next reference and rearrange those references to insert a new node, we are ready to address removal of nodes. We will continue to provide the root node and a position to the function. Also, because we might choose to remove the element at the 0 position, we will continue to return the root node of the resulting list. As with the **insertAtPosition**, we assume that the value for position is valid for this list.

```
function removeAtPosition(rootNode, position)
    if position equals 0
        return rootNode.Next
    else
        # follow Next until remove position is reached
        set n to rootNode
        set i to 1
        while i < position
            set n to n.Next
            set i to i + 1
        set n.Next to n.Next.Next
        return rootNode
```

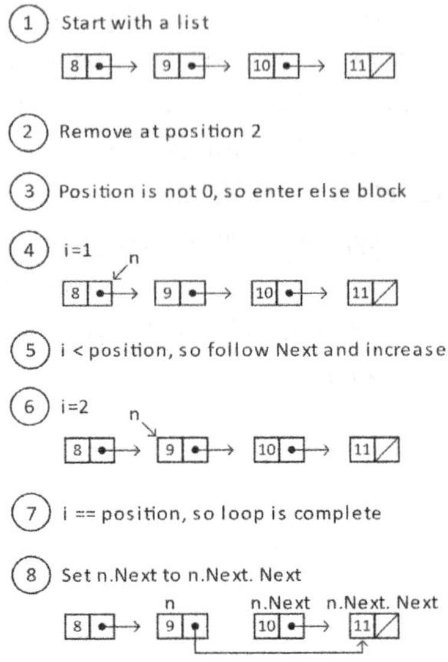

Figure 5.5

The result of the diagram above is that a traversal of the linked list will indeed include the values 8, 9, and 11 as desired. We should pay close attention to the node with value 10. Depending on the language we choose to implement linked lists, we may or may not be required to address this removed node. If the language's runtime is memory managed, you may simply ignore the unreferenced node. In languages that rely on manual memory management, you should be prepared to deallocate the storage. The runtime for the algorithm, as a function of the length of the list, is still O(n), with the special case of position as 0 running again in constant time.

Doubly Linked Lists

Consider a scenario where we want to track the sequence of changes to a shared document. Compared to arrays, a linked list can grow as needed and is better suited for the task. We choose to inspect the change at position 500 in the linked list at a cost of 499 dereferences. We then realize we stepped two changes too far. We are actually concerned with change 498. We must then incur a cost of 497 dereferences to simply move backward two steps. The issue is that our nodes currently only point to the next value in the sequence and not the previous. Luckily, we can simply choose to include a **Prior** reference.

```
class Node
    Integer Value
    Node Prior
    Node Next
```

Figure 5.6

The choice to track prior nodes in addition to the next nodes does come with trade-offs. First, the size of each node stored is now larger. If we assume an integer is the same size as a reference, we have likely increased the size of each node stored by 50%. Depending on the needs of the application, constraints of the physical device, and size of the linked list, this increase may or may not be acceptable.

We also have more complicated (and technically slightly slower) functions for insertion and removal of nodes. See the pseudocode below for considerations in the general case. The case when position is 0 has also been omitted. Completing that case is an exercise at the end of this chapter.

Reference Reassignment for Singly Linked List Insert

```
set newNode.Next to n.Next
set n.Next to newNode
```

Reference Reassignment for Doubly Linked List Insert

```
1 set newNode.Next to n.Next
2 set newNode.Prior to n
3 set n.Next to newNode
4 if newNode.Next is not null
5     set newNode.Next.Prior to newNode
```

If we would like to make use of a **Prior** reference, we now must maintain that value on each insertion and removal. At times it is as easy as setting a value on an object (line 2). Other times we have to introduce a new condition (line 4). In this case, we check **newNode.Next** against null because we may be inserting our new value at the end of the list, in which case, there will not be any node with **Prior** set to **newNode**. Doubly linked list insert now requires as many as 5 operations where we only had 2 for singly linked lists. While this does mean that doubly linked list insert is technically slower, we only perform these operations once per function call. As a result, we have two functions that run in O(n) even though one is technically faster than the other.

Returning to the change tracking example at the beginning of this section, we now have a means of moving forward and backward through our list of changes. If we wish to start our analysis of the change log at entry 500, it will indeed cost us 499 dereferences to reach that node. However, once at that node, we can inspect entry 498 with a cost of 2 dereferences by following the **Prior** reference.

Augmenting Linked Lists

Just as we saw when inserting or removing at position 0, we can often find clever ways to improve certain behaviors of linked lists. These improvements may lead to better runtimes or simply have a clearer intent. In this section, we will consider the most common ways linked lists are augmented. Generally speaking, we will follow the same strategy used for doubly linked lists. The main principle is this: At the time that we know some useful bit of information, we will choose to simply save it for later. This will lead to a marginally higher cost for certain operations and a larger amount of data to store, but certain operations will become much faster.

So far in this chapter, we have implicitly defined a list as a single node representing the root element. To augment our data structure, we will now more formally define the full concept of a list as follows. For the definition of this class, we can choose to use a singly or doubly linked node. For simplicity, examples in this section return to using a singly linked node.

```
class List
    Node Head   # the first element of the list
    Node Tail   # the last element of the list
    Integer Length
```

As was the case with doubly linked lists, our insert and remove code is now substantially more complicated. One nice benefit is that we now modify the list object and no longer need to return the root node.

```
1  function insertAtPosition(list, position, value)
2      # create a node to hold the new value
3      set newNode to an instance of Node
4      set newNode.Value to value
5
6      if list.Length equals 0
7          set list.Head to newNode
8          set list.Tail to newNode
9      else if position equals 0
10         set newNode.Next to list.Head
11         set list.Head to newNode
12     else
13         # follow Next until in position to insert
14         set n to list.Head
15         set i to 1
16         while i < position
17             set n to n.Next
18             set i to i + 1
19         set newNode.Next to n.Next
20         set n.Next to newNode
21         if position equals list.Length-1
22             set list.Tail to newNode
23
24     set list.Length to list.Length + 1
```

For each insert into the list, we must now maintain some new values on the list object. Lines 7, 8, 11, and 22 help keep track of when we have changed the head or tail of the list. Line 24 runs regardless of where the value was inserted because we now have one more element in the list.

This extra work was not in vain. Consider what was required if we wanted to write a function to return the last element of a linked list represented by the root node compared to running it on a list object.

Last Element Using Root Node and Next References

```
function lastElement(rootNode) # runs in O(n) time
    set n to rootNode
    while n.Next is not null
        set n to n.Next
    return n.Value
```

Last Element Using the List and Tail References

```
function lastElement(list) # runs in O(1) time
    return list.Tail.Value
```

The same improvements can be seen in retrieving the length of a list.

Length Using Root Node and Next References

```
function length(rootNode) # runs in O(n) time
    set n to rootNode
    set c to 0
    while n is not null
        set n to n.Next
        set c to c + 1
    return c
```

Length Using the List and Length Values

```
function length(list) # runs in O(1) time
    return list.Length
```

Abstract Data Types

Before closing this chapter on linked lists, we benefit from considering abstractions. An abstract data type (ADT) is a collection of operations we want to perform on a given data type. Just as we can imagine numerous implementations for a given data structure (maybe we change a for-loop to a while-loop or recursion), we can also imagine numerous data structures that satisfy an ADT.

Consider the operations defined above for linked lists. We often want to insert, delete, count, and iterate over list elements. We call this set of operations a list ADT. A list may be implemented using linked nodes, an array, or some other means. However, it must provide these four operations. Implied in this description of ADTs is the fact that we cannot discuss the asymptotic runtime or space requirements of an ADT. Without knowing how the ADT is implemented, we cannot conclude much (if anything) about the runtime. For example, we could conclude that iteration is no better than O(n) because iteration requires us to touch each element regardless of implementation. We could not determine anything about the runtime of lookup because an array would be O(1), a linked list O(n), and a skip list O(log n). This last data structure is not formally covered in this chapter.

In subsequent chapters, we will at times refer to ADTs without specifying the precise data structure. In doing so, we will be able to focus on the new data structure without concern for how the ADT is implemented. Naturally, when we address the runtimes and space utilization of those algorithms, we must choose between data structures.

Exercises

1. Write three functions that print all values in a singly linked list. Write one using each of the following: for-loop, while-loop, and recursion.

2. Write a **removeAtPosition** function for a doubly linked list that correctly maintains the **Prior** reference when the removal occurs at position 0, length – 1, or some arbitrary position in between.

3. Write a **removeAtPosition** function for a singly linked list that correctly maintains the **Head** and **Tail** references when the removal occurs at position 0, length – 1, or some arbitrary position between.

References

Cormen, Thomas H., Charles E. Leiserson, Ronald L. Rivest, and Clifford Stein. *Introduction to Algorithms*, 2nd ed. Cambridge, MA: The MIT Press, 2001.

6. Stacks and Queues

Learning Objectives

After reading this chapter you will...

- learn how data structures can help facilitate the orderly processing of organized data.
- learn the basic operations of stacks and queues.
- learn how abstract data types help us define concepts while delaying implementation details.

Introduction

Consider reading an article on Wikipedia. You read until you encounter an unfamiliar term, then you click it to open a new article. You continue this process until you have several pages open simultaneously. Then you receive a phone call. It is a close friend, and you choose to derail your studies for a few minutes. After your chat, you return to your browser and try to pick up where you left off. You find a series of tabs open. The tab farthest to the left was your original article. The tab farthest to the right was the most recent. The tabs between occur in the order in which you clicked related articles.

You decide you must be as systematic as possible as you get

back to your train of thought. Should you start at the leftmost or rightmost tab? What can we determine about these options?

- Left-to-right: In this strategy, you choose to go back to reading the original article you opened. After all, it was the most important when you sat down to start your research, so you should probably get started there. Then as needed, you can continue to work through other tabs. If you choose this strategy, it is because you assume that you had opened tabs because you might be interested in them later.
- Right-to-left: In this strategy, you assume that you opened "Formal system" because you wanted to understand it better while reading "Decidability (logic)," which was open because you were reading "Decision problem." If you choose this strategy, it is because you assume that you opened each additional tab due to some context present in the prior tab.

Both strategies share some commonalities. New information was encountered in some sort of linear fashion. The choice is which to prioritize. Is it important to process the information in the order it was received? Or is it more important to process the most recent information first? Here we have a problem with two reasonable solutions depending on our original expectations and desired outcome.

These options reflect two related but distinct abstractions. Both queues and stacks store values in a sequential manner in the order in which they arrived. A queue processes the input that arrived first before that which arrived second. For this reason, we consider queues to have a first-in-first-out (or FIFO) property. A stack instead processes the most recently received input first. Likewise, we consider stacks to have a last-in-first-out (or LIFO) property.

Queues are always present in our natural life. Whenever we "line up" for something, we have formed a queue. Queues have a visceral fairness about them. Those who arrived at the grocery store checkout first are served first. Those in line earlier for that big movie premiere are more likely to get seats. Queues are also convenient because they preserve a temporal sequence. Consider a system like a document editor. You specify changes to the document, and those changes are made. These must be processed in the order they are received because one change is likely dependent on whether the prior has been completed.

However, queues are not always the most expressive way to process a stream of data. Consider the call stack from our prior investigation of recursion in chapter 2. Each new function call encountered is processed within the context of the function call that came prior. While it is true that some other function call came before that, we are more concerned with maintaining the contextual reference rather than the temporal one. Stacks are also convenient when no implied order or precedence is required. In that case, inserting and retrieving data from the same end of the structure (LIFO) is often more efficient than the FIFO alternative.

You may have noticed that we have not explicitly referred to stacks and queues as data structures. While the term "data structure" may be considered appropriate for queues and stacks, a more appropriate description would be abstract data type. This distinction will be the first topic of this chapter.

Abstract Data Types

With linked lists, we described desired operations alongside a specific structure that satisfies that behavior. At times it is beneficial to decouple the two. The nodes represented the data structure, while the set of desired operations (such as insertion and deletion) formed what is called the abstract data type (ADT).

In this chapter, we define queues and stacks as abstract data types before specifying underlying data structures. Different reference texts specify slightly different operations. For the

purposes of this text, we define queue and stack with three defined operations:
Queue

- enqueue: places an element at the tail of the queue
- dequeue: removes the next available element from the head of the queue
- isEmpty: returns true or false depending on whether the underlying data structure has any remaining elements

Stack

- push: places an element at the top of a stack
- pop: removes an element from the top of a stack
- isEmpty: returns true or false depending on whether the underlying data structure has any remaining elements

We have three important nuances thus far:

1. We have not specified exactly how we intend to guarantee these operations or how we will store elements until we dequeue or pop them.
2. The above definitions use words like head, tail, and top. From one text to another, these words may change. The important part is that queues have elements entering on one end and leaving from the other. Stacks have elements entering and leaving from the same orientation.
3. We cannot yet specify the runtime of any of these operations. We can only do so once the actual underlying data structures have been specified.

In the following sections, we illustrate how these operations can be satisfied with linked lists or arrays.

Linked Lists

Linked lists are well suited for both queues and stacks. First, adding and removing elements is simply a matter of creating a node and setting a reference or removing a reference. Second, we have a fairly intuitive means of tracking both ends of the data structure. For purposes of this section, we will assume a singly linked node discussed earlier in this book.

```
class QueueList
    Node head
    Node tail

    function enqueue(value)
        set newNode to new instance of a node
        set newNode.Value to value
        if head is null
            set head to newNode
            set tail to newNode
        else
            set Tail.Next to newNode

    function dequeue()
        set value to head.Value
        set head to head.Next
        return value

    function isEmpty()
        return head equals null
```

```
class StackList
    Node top

    function push(value)
        set newNode to new instance of a node
        set newNode.Value to value
        if top is null
            set top to newNode
        else
            set newNode.Next to top
            set top to newNode

    function pop()
        set value to top.Value
        set top to top.Next
        return value

    function isEmpty()
        return top equals null
```

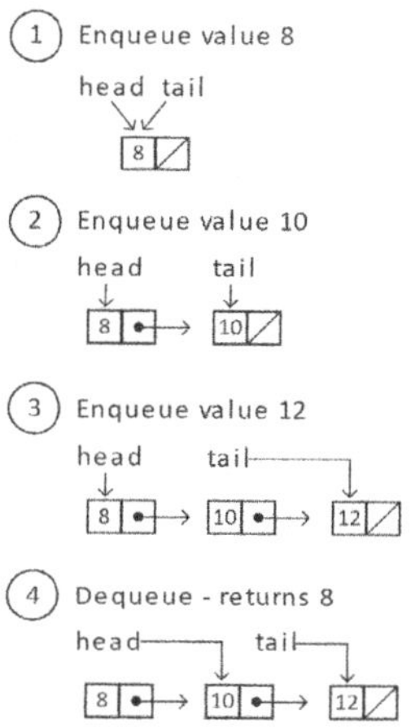

Figure 6.1

First, assess the runtime of each operation. Most operations on linked lists are O(n) due to the fact we have to traverse lists one node reference at a time. In this special case of queues and stacks, no traversal is necessary as long as we maintain a reference to the head or tail. No matter how many elements we enqueue or push, we only ever care about one or two nodes at any given time. The result is an elegant solution that runs in O(1) time for most operations.

Our analysis is not yet complete though, as we must also consider space. Considering that we are working with singly linked nodes, each has a reference and a value. If we want to store 100 integers in our queue or stack, we must have a node for each. Again, assuming that a reference is the same size as an integer, we result in a data structure with twice the overall storage footprint of the data itself. We do, however, still have the slight benefit that it does not have to be stored in contiguous memory locations.

Arrays

Just as we chose to implement a queue and a stack with a linked list, we could likewise choose to store the data in an array. As long as we have definitions for all operations that satisfy the operations specified for the ADT, the result will be just as effective.

Before defining the data structure, we should address the primary challenge when creating array-based queues and stacks. Arrays have a fixed length. Operations on queues and stacks are primarily about adding or removing elements from the underlying structure. How can we use a fixed-length data structure to implement an ADT that necessitates change? The answer typically involves some clever tricks. We define the data structures below and then systematically walk through these tricks:

```
class QueueArray
    Array data
    Integer headPos
    Integer tailPos
    set headPos to 0
    set tailPos to -1

    function enqueue(value)
        if tailPos + 1 equals length(data)
            Copy values into larger array D
            set data to D
        set tailPos to tailPos + 1
        set data[tailPos] to value

    function dequeue()
        set value to data[headPos]
        set headPos to headPos+1
        return value

    function isEmpty()
        return tailPos < headPos
```

```
class StackArray
    Array data
    Integer topPos
    set topPos to -1

    function push(value)
        if topPos + 1 equals length(data)
            Copy values into larger array D
            set data to D
        set topPos to topPos + 1
        set data[topPos] to value

    function pop()
        set value to data[topPos]
        set topPos to topPos - 1
        return value

    function isEmpty()
        return topPos < 0
```

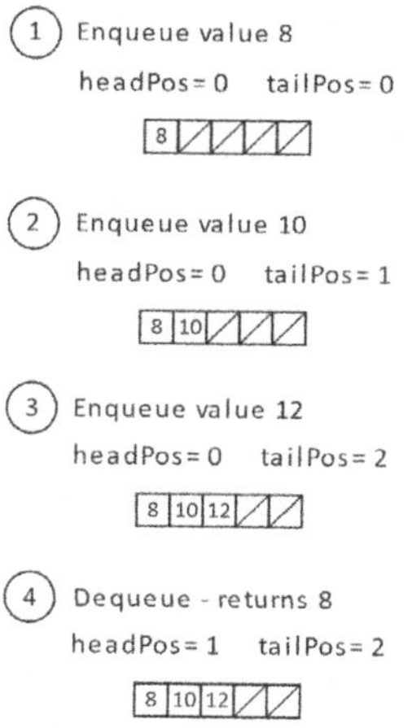

Figure 6.2

Storage Allocation Considerations

How can we ensure we have enough space in our array for the next invocation of enqueue or push? The typical strategy is to initially allocate an array with some amount of excess capacity then reallocate and copy to a larger array as needed. In the **QueueArray** and **StackArray** classes, these reallocations occur at lines 11 and 9. This challenge is a balancing act. On one hand, allocating too much space results in waste, as it cannot be used for other purposes while it is part of the allocated array. On the other hand, allocating too little results in more frequent reallocations. This creates issues in two ways.

First, when reallocation is required, our constant-time enqueue or push operation momentarily becomes O(n) with regard to the number of elements stored. From a big-picture perspective, this is not a major issue. Recall from chapter 1 that infrequently required additional steps permit us to talk in terms of amortized cost. The reallocation is indeed expensive but happens so infrequently that other enqueue or push invocations essentially absorb that cost. The result is in an amortized runtime of O(1) for these two operations.

Amortized runtimes do not always tell the same story as nonamortized. In both cases, we can generally expect an invocation to complete efficiently regardless of the number of elements in the queue or stack. The key word here is generally. Consider an application that is responsible for processing a queue with large volumes of real-time data. It may successfully invoke enqueue hundreds of times with a runtime between 5 and 15 milliseconds. A problem arises when the application is dependent on this level of performance because, at some point, an enqueue might trigger a reallocation and instead complete in 500 milliseconds.

The second issue with reallocations pertains to space usage. Not only are we regularly wasting space, but the reallocation itself temporarily consumes excess memory. If we are reallocating and copying from one large array to another, there is a small window of time where the original array and the new array are both consuming precious contiguous blocks of memory. These blocks can be expensive to allocate and even create failures in cases where a large enough contiguous block is not available.

Another clever trick can be applied to queues. Consider what happens when we have an initial array of length 10. We then enqueue and dequeue 10 elements, shifting our head and tail indexes to the end of the array. Our next enqueue will trigger a reallocation even though we are not currently using all 10 spots in our current array. We have the required space allocated but no clear means to access it. The trick lies in how we index into the array. Instead of using **tailPos** and **headPos** alone, we still use them but modulo the size of the array. If we do so, our eleventh enqueue and dequeue will use index 10. The result of 10 modulo 10 is 0 and will result in the usage of array position 0. The twelfth enqueue will use index 11, which becomes 1 after modulo 10. Following this strategy (often referred to as wrapping around), we can continue to reuse the original space allocated if our dequeue rate does not lag behind our enqueue rate.

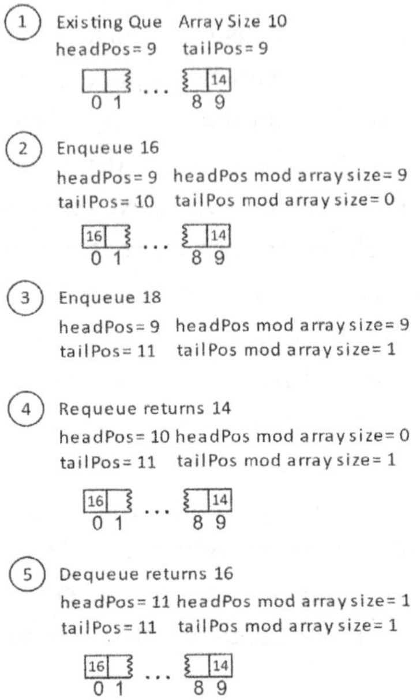

Figure 6.3

Practical Considerations

At the end of the last section, we devised a clever solution to improve the utilization of space already allocated. Although it has improved the data structure on the whole, it now creates a new issue. We now need a new means of determining when to reallocate the array. This can be addressed by keeping an explicit count of how many array elements are currently being used. However, once we have solved that issue, we must then decide how to handle **headPos** and **tailPos**. We had relied on using them modulo the size of the array, but now that size has changed. This is the nature of data structures. We often perceive them as static constructs that we study, memorize, and reimplement in new languages. In reality, they are dynamic and evolve as we apply basic concepts to novel problems.

It is also often the case that ADTs and data structures do not exist in isolation. Many languages blend queues and stacks with lists into a single data type. C++ vectors, JavaScript arrays, and Python lists all implement certain operations of these three ADTs (queues, stacks, and lists).

Again, specifying operations in an ADT does not necessarily imply any underlying structure. How, then, do these languages store sequential data? The process typically follows a narrative similar to that of wrapping around in array-based queues. A developer desires some traits of a clever solution, but implementing such a solution then leads to a new set of challenges. Working through these challenges requires both in-depth theoretical knowledge and a grasp of the real-world system. Once considering that real-world context, where modern memory is abundant and constant-time lookups are strongly desirable, most sequential data types make heavier use of arrays than linked lists.

Exercises

1.
In your choice of high-level language, implement stack and queue with a built-in sequential data type (such as List for Python or Vector for C++). What operations did you use on that data type? Research your language's documentation to determine the runtime of your enqueue, dequeue, push, and pop operations.

2.
The wraparound trick applied to array-based queues

only works if the application dequeues at least as fast as it enqueues. Describe real-world scenarios where we expect this might be true.

References

Cormen, Thomas H., Charles E. Leiserson, Ronald L. Rivest, and Clifford Stein. *Introduction to Algorithms*, 2nd ed. Cambridge, MA: The MIT Press, 2001.

7. Hashing and Hash Tables

Learning Objectives

After reading this chapter you will...

- understand what hash functions are and what they do.
- be able to use hash functions to implement an efficient search data structure, a hash table.
- understand the open addressing strategy for implementing hash tables.
- understand the potential problems with using hash functions for searching.
- be able to implement a hash table using data structure composition and the separate chaining strategy.

Introduction

In chapter 4, we explored a very important class of problems: searching. As the number of items increases, Linear Search becomes very costly, as every search costs O(n). Additionally, the real-world time cost of searching increases when the number of searches (also called queries) increases. For this reason, we explored sorting our "keys" (our unique identifiers) and then using Binary Search on data sets that get searched a lot. Binary Search improves our performance to O(log n) for searches. In this chapter, we explore ways to further improve search to approximately O(1) or constant time on average. There are some considerations, but with a good design, searches can be made extremely efficient. The key to this seemingly magic algorithm is the hash function. Let's explore hashes a bit more in-depth.

Hash Functions

If you have ever eaten breakfast at a diner in the USA, you were probably served some hash browns. These are potatoes that have been very finely chopped up and cooked. In fact, this is where the "hash" of the hash function gets its name. A hash function takes a number, the key, and generates a new number using information in the original key value. So at some level, it takes information stored in the key, chops the bits or digits into parts, then rearranges or combines them to generate a new number. The important part, though, is that the hash function will always generate the same output number given the input key. There are many different types of hash functions. Let's look at a simple one that hashes the key 137. We will use a small subscript of 2 when indicating binary numbers.

$137_{10} = 1000\ 1001_2$

Chop the binary number in half and convert to decimal.

$1000^2 = 8$

$1001^2 = 9$

Add each part to generate another number

$8 + 9 = 17$

Figure 7.1

We can generate hashes using strings or text as well. We can extract letters from the string, convert them to numbers, and add them together. Here is an example of a different hash function that processes a string:

Key = "Samantha"

Extract the first and last letter from the string
Convert them to integers using ASCII codes.
"S" = 83 in ASCII
"a" = 97 in ASCII

Add the integers together
83 + 97 = 180

Figure 7.2

There are many hash functions that can be constructed for keys. For our purposes, we want hash functions that exhibit some special properties. In this chapter, we will be constructing a lookup table using hashes. Suppose we wanted to store student data for 100 students. If our hash function could take the student's name as the key and generate a unique index for every student, we could store all their data in an array of objects and search using the hash. This would give us constant time, or O(1), lookups for any search! Students could have any name, which would be a vast set of possible keys. The hash function would look at the name and generate a valid array index in the range of 0 to 99.

The hash functions useful in this chapter map keys from

a very large domain into a small range represented by the size of the table or array we want to use. Generally, this cannot be done perfectly, and we get some "**collisions**" where different keys are hashed to the same index. This is one of the main problems we will try to fix in this chapter. So one property of the hash function we want is that it leads to few collisions. Since a perfect hash is difficult to achieve, we may settle for an unbiased one. A hash function is said to be **uniform** if it hashes keys to indexes with roughly equal probability. This means that if the keys are uniformly distributed, the generated hash values from the keys should also be roughly uniformly distributed. To state that another way, when considered over all k keys, the probability h(k) = a is approximately the same as the probability that h(k) = b. Even with a nice hash function, collisions can still happen. Let's explore how to tackle these problems.

Hash Tables

Once you have finished reading this chapter, you will understand the idea behind hash tables. A **hash table** is essentially a lookup table that allows extremely fast search operations. This data structure is also known as a hash map, associative array, or dictionary. Or more accurately, a hash table may be used to implement associative arrays and dictionaries. There may be some dispute on these exact terms, but the general idea is this: we want a data structure that associates a key and some value, and it must efficiently find the value when given the key. It may be helpful to think of a hash table as a generalization of an array where any data type can be used as an index. This is made possible by applying a hash function to the key value.

For this chapter, we will keep things simple and only use integer keys. Nearly all modern programming languages provide a built-in hash function or several hash functions. These language library-provided functions can hash nearly all data types. It is recommended that you use the language-provided hash functions in most cases. These are functions that generally have the nice properties we are looking for, and they usually support all data types as inputs (rather than just integers).

A Hash Table Using Open Addressing

Suppose we want to construct a fast-access database for a list of students. We will use the Student class from chapter 4. We will slightly alter the names though. For this example, we will use the variable name key rather than member_id to simplify the code and make the meaning a bit clearer.

```
class Student
    Integer key
    String name
    String email
```

We want our database data structure to be able to support searches using a search operation. Sometimes the term "find" is used rather than "search" for this operation. We will be consistent with chapter 4 and use the term "search" for this operation. As the database will be searched frequently, we want search to be very efficient. We also need some way to add and remove students from the database. This means our data structure should support the add and remove operations.

The first strategy we will explore with hash tables is known as **open addressing**. This means that we will allot some storage space in memory and place new data records into an open position addressed by the hash function. This is usually done using an array. Let the variable size represent the number of positions in the array.

With the size of our array known, we can introduce a simple hash function where mod is the modulo or remainder operator.

```
function hash(key)
    return key mod size
```

This hash function maps the key to a valid array index. This can be done in constant time, O(1). When searching for a student in our database, we could do something like this:

```
set index to hash(search_key)
return array[index]
```

This would ensure a constant-time search operation. There is a problem, though. Suppose our array had a size of 10. What would happen if we searched for the student with key 18 and another student with key 28? Well, 18 mod 10 is 8, and 28 mod 10 is 8. This simple approach tries to look for the same student in the same array address. This is known as a **collision** or a **hash collision**.

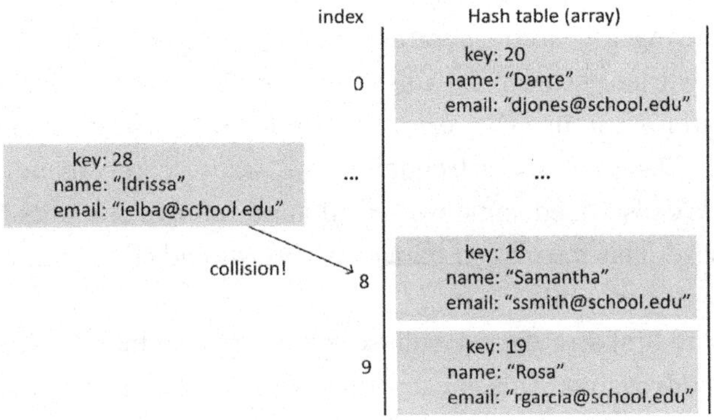

Figure 7.3

We have two options to deal with this problem. First, we could use a different hash function. There may be another way to hash the key to avoid these collisions. Some algorithms can calculate such a function, but they require knowledge of all the keys that will be used. So this is a difficult option most of the time. The second alternative would be to introduce some policy for dealing with these collisions. In this chapter, we will take the second approach and introduce a strategy known as **probing**. Probing tries to find the correct key by "probing" or checking other positions relative to the initial hashed address that resulted in the collision. Let's explore this idea with a more detailed example and implementation.

Open Addressing with Linear Probing

Let us begin by specifying our hash table data structure. This class will need a few class functions that we will specify below, but first let's give our hash table some data. Our table will need an array of Student records and a size variable.

```
class HashTable
    Student Array table
    Integer size
```

To add new students to our data structure, we will use an add function. There is a simple implementation for this function without probing. We will consider this approach and then improve on it. Assume that the add function belongs to the HashTable class, meaning that table and size are both accessible without passing them to the function.

```
function add(student)
    set index to hash(student.key)
    set table[index] to student
```

Once a student is added, the HashTable could find the student using the search function. We will have our search function return the index of the student in the array or -1 if the student cannot be found.

```
function search(key)
    set index to hash(key)
    if key equals table[index].key
        return index
    else
        return -1
```

This approach could work assuming our hash was perfect. This is usually not the case though. We will extend the class to handle collisions. First, let's explore an example of our probing strategy.

Probing

Suppose we try to insert a student, marked as "A," into the database and find that the student's hashed position is already occupied. In this example, student A is hashed to position 2, but we have a collision.

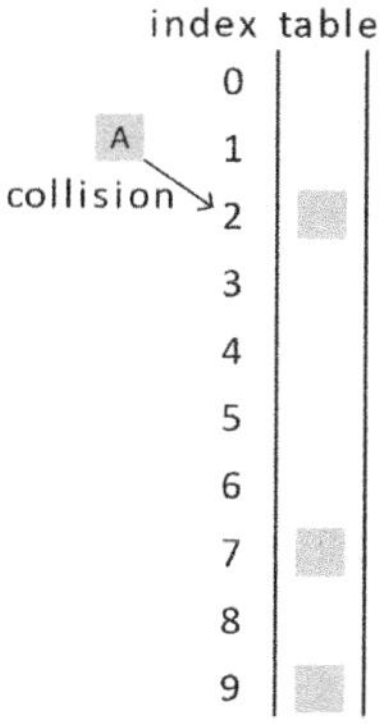

Figure 7.4

With probing, we would try the next position in the **probe sequence**. The probe sequence specifies which positions to try next. We will use a simple probe sequence known as **linear probing**. Linear probing will have us just try the next position in the array.

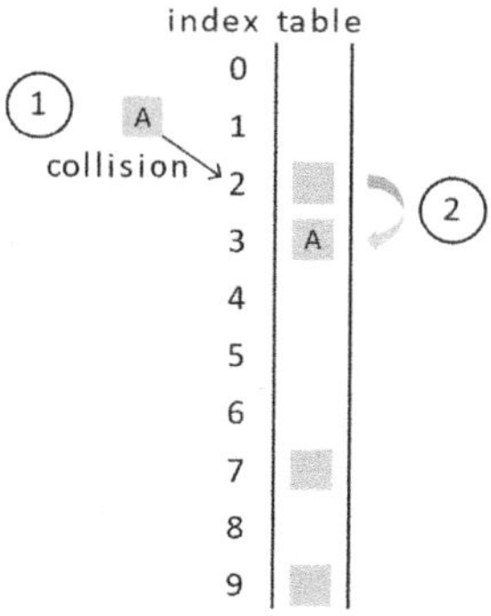

Figure 7.5

This figure shows that first we get a collision when trying to insert student A. Second, we probe the next position in the array and find that it is empty, so student A is inserted into this array slot. If another collision happens on the same hash position, linear probing has us continue to explore further into the array and away from the original hash position.

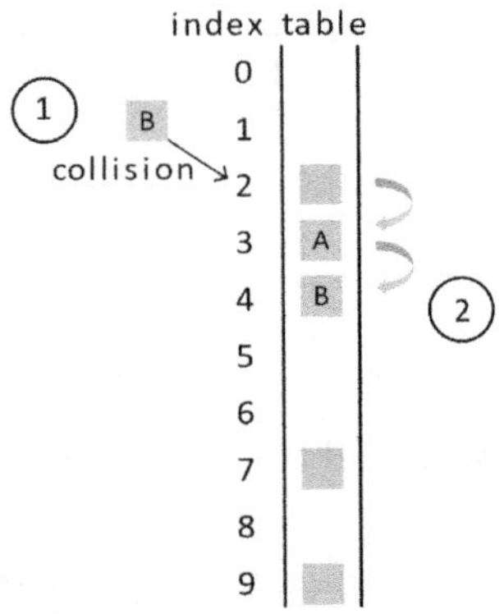

Figure 7.6

This figure shows that another collision will require more probing. You may now be thinking, "This could lead to trouble." You would be right. Using open addressing with probing means that collisions can start to cause a lot of problems. The frequency of collisions will quickly lead to poor performance. We will revisit this soon when we discuss time complexity. For now, we have a few other problems with this approach.

Add and Search with Probing

Let us tackle a relatively simple problem first. How can we

implement our probe sequence? We want our hash function to return the proper hash the first time it is used. If we have a collision, our hash needs to return the original value plus 1. If we have two collisions, we need the original value plus 2, and so on. For this, we will create a new hashing function that takes two input parameters.

```
function hash(key, collisions)
    set initialIndex to key mod size
    set index to initialIndex + collisions
    return index mod size
```

With this function, hash(2,2) would give the value 4 as in the previous figure. In that example, when trying to insert student B, we get an initial collision followed by a second collision with student A that was just inserted. Finally, student B is inserted into position 4.

Did you notice the other problem? How will we check to see if the space in the array is occupied? There are a variety of approaches to solving this problem. We will take a simple approach that uses a secondary array of status values. This array will be used to mark which table spaces are occupied and which are available. We will add an integer array called status to our data structure. This approach will simplify the code and prepare our HashTable to support remove (delete) operations. The new HashTable will be defined as follows:

```
class HashTable
    Student Array table
    Integer Array status
    Integer size
```

We will assign a status value of 0 to an empty slot and a value of 1 to an occupied slot. Now to check if a space is open and available, the code could just check to see if the status value at that index is 0. If the status is 1, the position is filled, and adding to that location results in a collision. Now let's use this information to correct our add function for using linear probing. We will assume that all the status values are initialized with 0 when the HashTable is constructed.

```
function add(student)
    set collisions to 0
    set index to hash(student.key, collisions)

    # continue to probe if the positions are occupied
    while status[index] equals 1 and collisions < size
        set collisions to collisions + 1
        set index to hash(student.key, collisions)

    # if the spot is available, add the student
    if status[index] is not 1
        set table[index] to student
        set status[index] to 1
    else
        # All positions must be occupied
        print "HashTable is full!"
```

Now that we can add students to the table, let us develop the search function to deal with collisions. The search function will be like add. For this algorithm, status[index] should be 1 inside the while-loop, but we will allow for −1 values a bit later. This is why 0 is not used here.

```
function search(key)
    set collisions to 0
    set index to hash(key, collisions)
    set found to False

    # continue to probe if the positions are occupied
    while status[index] is not 0 and found equals False and collisions < size
        if status[index] equals 1 and table[index].key equals key
            set found to True
        else
            set collisions to collisions + 1
            set index to hash(key, collisions)

    # if the keys match, return the array index
    if found equals True
        return index
    else
        return -1
```

We need to discuss the last operation now: remove. The remove operation may also be called delete in some contexts. The meaning is the same though. We want a way to remove students from the database. Let's think about what happens when a student is removed from the database. Think back to the collision example where student B is inserted into the database. What would happen if A was removed and then we searched for B?

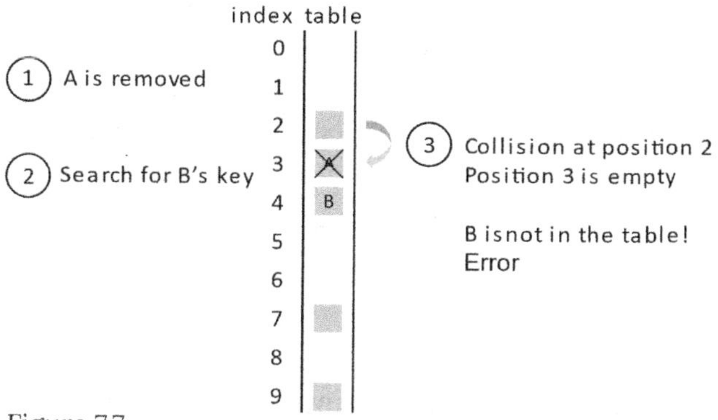

Figure 7.7

If we just marked a position as open after a remove operation, we would get an error like the one illustrated above. With this sequence of steps, it seems like B is not in the table because we found an open position as we searched for it. We need to deal with this problem. Luckily, we have laid the foundation for a simple solution. Rather than marking a deleted slot as open, we will give it a deleted status code. In our status array, any value of -1 will indicate that a student was deleted from the table. This will solve the problem above by allowing searches to proceed past these deleted positions in the probe sequence.

The following function can be used to implement the remove function. This approach relies on our search function that returns the correct index. Notice how the status array is updated.

```
function remove(key)
    set index to search(key)
    # if the key is found, update its status to -1, "deleted"
    if index is not -1
        set status[index] to -1
```

Depending on your implementation, you may also want to free the memory at table[index] at line 5. We are assuming that student records are stored directly in the array and will be overwritten on the next add operation for that position. If references are used, freeing the data may need to be explicit.

Take a careful look back at the search function to convince yourself that this is correct. When the status is −1, the search function should proceed through past collisions to work correctly. We now have a correct implementation of a hash table. There are some serious drawbacks though. Let us now discuss performance concerns with our hash table.

Complexity and Performance

We saw that adding more students to the hash table can lead to collisions. When we have collisions, the probing sequence places the colliding student near the original student record. Think about the situation below that builds off one of our previous examples:

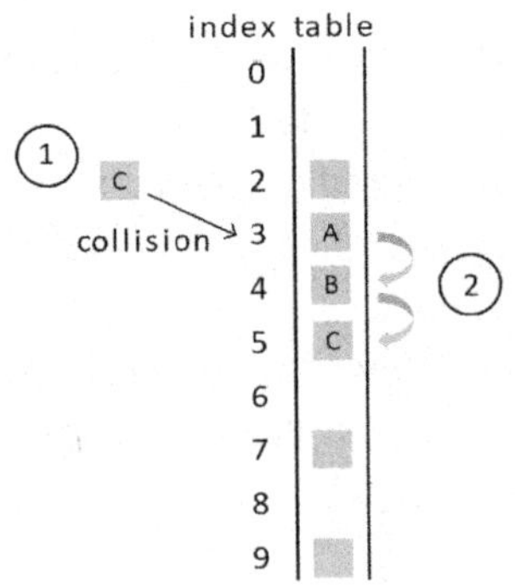

Figure 7.8

Suppose that we try to add student C to the table and C's key hashes to the index 3. No other student's key hashes to position 3, but we still get 2 collisions. This clump of records is known as a **cluster**. You can see that a few collisions lead to more collisions and the clusters start to grow and grow. In this example, collisions now result if we get keys that hash to any index between 2 and 5.

What does this mean? Well, if the table is mostly empty and our hash function does a decent job of avoiding collisions, then add and search should both be very efficient. We may have a few collisions, but our probe sequences would be short and on the order of a constant number of operations. As the table fills up, we get some collisions and some clusters. Then with clustering, we get more collisions and more clustering as a result. Now our searches are taking many more operations, and they may approach O(n) especially when the table is full and our search key is not actually in the database. We will explore this in a bit more detail.

A load factor is introduced to quantify how full or empty the table is. This is usually denoted as α or the Greek lowercase *alpha*. We will just use an uppercase L. The load factor can be defined as simply the ratio of added elements to the total capacity. In our table, the capacity is represented by the size variable. Let n be the number of elements to be added to the database. Then the overall load factor for the hash table would be L = n / size. For our table, L must be less than 1, as we can only store as many students as we have space in the array.

How does this relate to runtime complexity? Well, in the strict sense, the worst-case performance for searches would be O(n). This is represented by the fact that when the table is full, we must check nearly all positions in the table. On the other hand, our analysis of Quick Sort showed that the expected worst-case performance can mean we get a very efficient and highly useful algorithm even if some cases may be problematic. This is the case with hash tables. Our main interest is in the average case performance and understanding how to avoid the worst-case situation. This is where the load factor comes into play. Donald

Knuth is credited with calculating the average number of probes needed for linear probing in both a **successful search** and the more expensive **unsuccessful search**. Here, a successful search means that the item is found in the table. An unsuccessful search means the item was searched for but not found to be in the table. These search cost values depend on the L value. This makes sense, as a mostly empty table will be easy to insert into and yield few collisions.

The expected number of probes for a **successful search** with linear probing is as follows:

$$\frac{1}{2}\left\{1 + \frac{1}{(1-L)}\right\}.$$

For **unsuccessful searches**, the number of probes is larger:

$$\frac{1}{2}\left\{1 + \frac{1}{(1-L)^2}\right\}.$$

Let's put these values in context. Suppose our table size is 50 and there are 10 student records inserted into the table giving a load factor of 10/50 = 0.2. This means on average a successful search needs 1.125 probes. If the table instead contains 45 students, we can expect an average of 5.5 probes with an L of 45/50 = 0.9. This is the average. Some may take longer. The unsuccessful search yields even worse results. With an L of 10/50 = 0.2, an unsuccessful search would yield an average of 1.28 probes. With a table of lead L = 45/50 = 0.9, the average number of probes would be 50.5. This is close to the worst-case O(n) performance.

We can see that the average complexity is heavily influenced by the load factor L. This is true of all open addressing hash table methods. For this reason, many hash table data structures will detect that the load is high and then dynamically reallocate a larger array for the data. This increases capacity and reduces the load factor. This approach is also helpful when the table accumulates a lot of deleted entries. We will revisit this idea later in the chapter. Although linear probing has some poor performance at high loads, the nature of checking local positions has some advantages with processor caches. This is another important idea that makes linear probing very efficient in practice.

The space complexity of a hash table should be clear. We need enough space to store the elements; therefore, the space complexity is O(n). This is true of all the open addressing methods.

Other Probing Strategies

One major problem with linear probing is that as collisions occur, clusters begin to grow and grow. This blocks other hash positions and leads to more collisions and therefore more clustering. One strategy to reduce the cluster growth is to use a different probing sequence. In this section, we will look at two popular alternatives to linear probing. These are the methods of **quadratic probing** and

double hashing. Thanks to the design of our **HashTable** in the previous section, we can simply define new hash functions. This modular design means that changing the functionality of the data structure can be done by changing a single function. This kind of design is sometimes difficult to achieve, but it can greatly reduce repeated code.

Quadratic Probing

One alternative to linear probing is quadratic probing. This approach generates a probe sequence that increases by the square of the number of collisions. One simple form of quadratic probing could be implemented as follows:

```
function hash(key, collisions)
    set initialIndex to key mod size
    set index to initialIndex + collisions * collisions
    return index mod size
```

The following illustration shows how this might improve on the problem of clustering we saw in the section on linear probing:

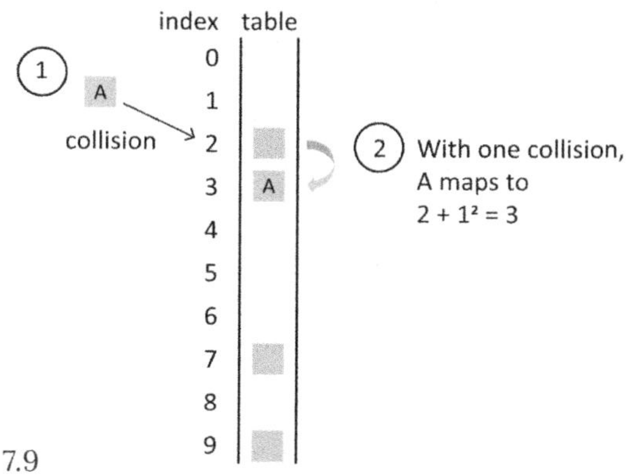

Figure 7.9

With one collision, student A still maps to position 3 because $2 + 1^2 = 3$. When B is mapped though, it results in 2 collisions. Ultimately, it lands in position 6 because $2 + 2^2 = 6$, as the following figure shows:

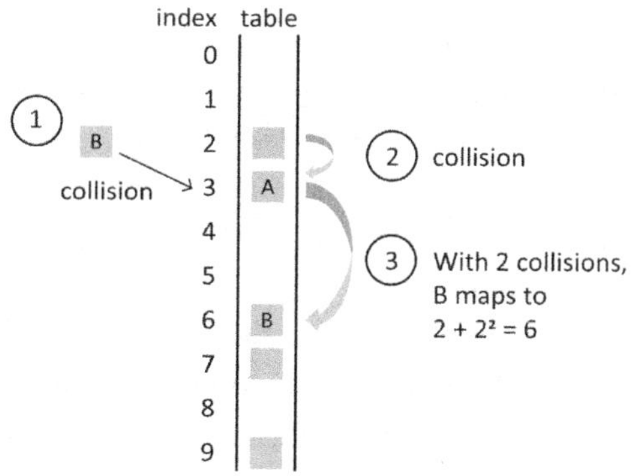

Figure 7.10

When student C is added, it will land in position 4, as $3 + 1^2 = 4$. The following figure shows this situation:

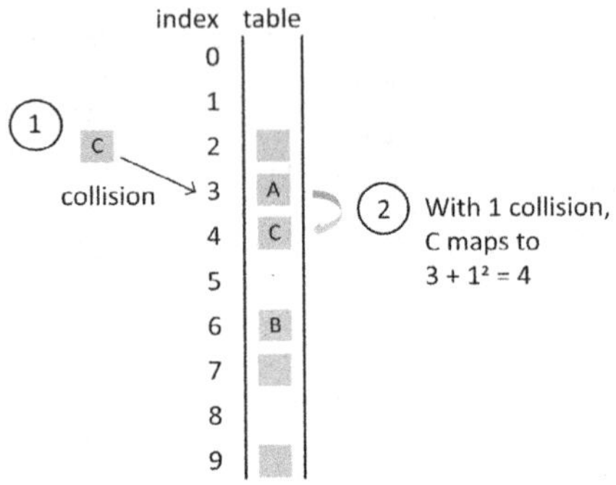

Figure 7.11

Now instead of one large primary cluster, we have two somewhat smaller clusters. While quadratic probing reduces the problems associated with primary clustering, it leads to **secondary clustering**.

One other problem with quadratic probing comes from the probe sequence. Using the approach we showed where the hash is calculated using a formula like $h(k) + c^2$, we will only use about size/2 possible indexes. Look at the following sequence: 1, 4, 9, 16, 25, 36, 49, 64, 81, 100, 121, 144. Now think about taking these values after applying **mod** 10. We get 1, 4, 9, 6, 5, 6, 9, 4, 1, 0, 1, 4. These give only 6 unique values. The same behavior is seen for any mod value or table size. For this reason, quadratic probing usually terminates once the number of collisions is half of the table size. We can make this modification to our algorithm by modifying the probing loop in the add and search functions.

For the add function, we would use

```
# quadratic probing loop:  add function
# continue to probe if the positions are occupied
while status[index] equals 1 and collisions < size/2
    set collisions to collisions + 1
    set index to hash(student.key, collisions)
```

For the search function, we would use

```
# quadratic probing loop:  search function
# continue to probe if the positions are occupied
while status[index] is not 0 and found equals False and collisions < size/2
    if status[index] equals 1 and table[index].key equals key
        set found to True
    else
        set collisions to collisions + 1
        set index to hash(key, collisions)
```

When adding, it is assumed that encountering size/2 collisions means that the table is full. It is possible that this is incorrect. There may be open positions available even after quadratic probing has failed. If attempting to add fails, it is a good indicator that the load factor has become too high anyway, and the table needs to be expanded and rebuilt.

Double Hashing

In this section, we will look at an implementation of a hashing collision strategy that approaches the ideal strategy for an open addressed hash table. We will also discuss how to choose a good table size such that our hash functions perform better when our keys do not follow a random uniform distribution.

Choosing a Table Size

So far, we have chosen a table size of 10 in our examples. This has made it easy to think about what hash value is generated from a base-10 numerical key. This would be fine assuming our key distribution was truly uniform in the key domain. In practice, keys can have some properties that result in biases and ultimately nonuniform distributions. Take, for example, the use of a memory address as a key. On many computer systems, memory addresses are multiples of 4. As another example, in English, the letter "e" is far more common than other letters. This might result in keys generated from ASCII text having a nonuniform distribution.

Let's look at an example of when this can become a problem. Suppose we have a table of size 12 and our keys are all multiples of 4. This would result in all keys being initially hashed to only the indexes 0, 4, or 8. For both linear probing and quadratic probing, any key with the initial hash value will give the same probing sequence. So this example gives an especially bad situation resulting in poor performance under both linear probing and quadratic probing. Now suppose that we used a prime number rather than 12, such as 13. The table below gives a sequence of multiples of 4 and the resulting mod values when divided by 12 and 13.

key	keymod12	keymod13
4	4	4
8	8	8
12	0	12
16	4	3
20	8	7
24	0	11
28	4	2

Figure 7.12

It is easy to see that using 13 performs much better than 12. In general, it is favored to use a table size that is a prime value. The approach of using a prime number in hash-based indexing is credited to Arnold Dumey in a 1956 work. This helps with nonuniform key distributions.

Implementing Double Hashing

As the name implies, double hashing uses two hash functions rather than one. Let's look at the specific problem this addresses. Suppose we are using the good practice of having size be a prime number. This still cannot overcome the problem in probing methods of having the same initial hash index. Consider the following situation. Suppose k_1 is 13 and k_2 is 26. Both keys will generate a hashed value of 0 using mod 13. The probing sequence for k_1 in linear probing is this:

$h(k_1,0) = 0$, $h(k_1,1) = 1$, $h(k_1,2) = 2$, and so on. The same is true for k_2.

Quadratic probing has the same problem:

$h_Q(k_1, 0) = 0$, $h_Q(k_1, 1) = 1$, $h_Q(k_1, 2) = 2$. This is the same probe sequence for k_2.

Let's walk through the quadratic probing sequence a little more carefully to make it clear. Recall that

$$h_Q(k,c) = (k \textbf{ mod } size + c^2) \textbf{ mod } size$$

using quadratic probing. The following table gives the probe sequence for $k_1 = 13$ and $k_2 = 26$ using quadratic probing:

collisions	h(13, c)	h(26, c)
0	(13 mod 13 + 0) mod 13 = 0	(26 mod 13 + 0) mod 13 = 0
1	(13 mod 13 + 1) mod 13 = 1	(26 mod 13 + 1) mod 13 = 1
2	(13 mod 13 + 4) mod 13 = 4	(26 mod 13 + 4) mod 13 = 4
3	(13 mod 13 + 9) mod 13 = 9	(26 mod 13 + 9) mod 13 = 9
4	(13 mod 13 + 16) mod 13 = 3	(26 mod 13 + 16) mod 13 = 3

Figure 7.13

The probe sequence is identical given the same initial hash. To solve this problem, double hashing was introduced. The idea is simple. A second hash function is introduced, and the probe sequence is generated by multiplying the number of collisions by a second hash function. How should we choose this second hash function? Well, it turns out that choosing a second prime number smaller than size works well in practice.

Let's create two hash functions $h_1(k)$ and $h_2(k)$. Now let p_1 be a prime number that is equal to size. Let p_2 be a prime number such that $p_2 < p_1$. We can now define our functions and the final double hash function:

$$h_1(k) = k \textbf{ mod } p_1$$

$$h_2(k) = k \textbf{ mod } p_2.$$

The final function to generate the probe sequence is here:

$$h(k, c) = (h_1(k) + c*h_2(k)) \textbf{ mod } size.$$

Let's let $p_1 = 13 = size$ and $p_2 = 11$ for our example. How would this change the probe sequence for our keys 13 and 26? In this case $h_1(13) = h_1(26) = 0$, but $h_2(13) = 2$, $h_2(26) = 4$.

Consider the following table:

collisions	h(13, c)	h(26, c)
0	(0 + 0*2) mod 13 = 0	(0 + 0*4) mod 13 = 0
1	(0 + 1*2) mod 13 = 2	(0 + 1*4) mod 13 = 4
2	(0 + 2*2) mod 13 = 4	(0 + 2*4) mod 13 = 8
3	(0 + 3*2) mod 13 = 6	(0 + 3*4) mod 13 = 12
4	(0 + 4*2) mod 13 = 8	(0 + 4*4) mod 13 = 3

Figure 7.14

Now that we understand double hashing, let's start to explore one implementation in code. We will create two hash functions as follows:

```
function hashOne(key)
    # assume size is a prime number
    return key mod size
```

The second hash function will use a variable called **prime**, which has a value that is a prime number smaller than size.

```
function hashTwo(key)
    # assume prime is less than size
    return key mod prime
```

Finally, our hash function with a collisions parameter is developed below:

```
function hash(key, collisions)
    set initialIndex to hashOne(key)
    set index to initialIndex + collisions * hashTwo(key)
    return index mod size
```

As before, these can be easily added to our HashTable data structure without changing much of the code. We would simply add the **hashOne** and **hashTwo** functions and replace the two-parameter hash function.

Complexity of Open Addressing Methods

Open addressing strategies for implementing hash tables that use probing all have some features in common. Generally speaking, they all require O(n) space to store the data entries. In the worst case, search-time cost could be as bad as O(n), where the data structure checks every entry for the correct key. This is not the full story though.

As we discussed before with linear probing, when a table is mostly empty, adding data or searching will be fast. First, check the position in O(1) with the hash. Next, if the key is not found and the table is mostly empty, we will check a small constant number of probes. Search and insert would be O(1), but only if it's mostly empty. The next question that comes to mind is "What does 'mostly empty' mean?" Well, we used a special value to quantify the "fullness" level of the table. We called this the load factor, which we represented with L.

Let's explore L and how it is used to reason about the average runtime complexity of open addressing hash tables. To better understand this idea, we will use an ideal model of open addressing with probing methods. This is known as uniform hashing, which was discussed a bit before. Remember the problems of linear probing and quadratic probing. If any value gives the same initial hash, they end up with the same probe sequence. This leads to clustering and degrading performance. Double hashing is a close approximation to uniform hashing. Let's consider double hashing with two distinct keys, k_1 and k_2. We saw that when $h_1(k_1) = h_1(k_2)$, we can still get a different probe sequence if $h_2(k_1) \neq h_2(k_2)$. An ideal scenario would be that every unique key generates a unique but uniform random sequence of probe indexes. This is known as uniform hashing. Under this model, thinking about the average number of probes in a search is a little easier. Let's think this through.

Remember that the load on the table is the ratio of filled to the total number of available positions in the table. If n elements have been inserted into the table, the load is L = n / size. Let's consider the case of an unsuccessful search. How many probes would we expect to make given that the load is L? We will make at least one check, but next, the probability that we would probe again would be L. Why? Well, if we found one of the (size – n) open positions, the search would have ended without probing. So the probability of one unsuccessful probe is L. What about the probability of two unsuccessful probes? The search would have failed in the first probe with probability L, and then it would fail again in trying to find one of the (n – 1) occupied positions among the (size – 1) remaining available positions. This leads to a probability of

$$P(2\,probes) = \left(\frac{n}{size}\right) \cdot \left(\frac{n-1}{size-1}\right) = L \cdot \left(\frac{n-1}{size-1}\right).$$

Things would progress from there. For 3 probes, we get the following:

$$P(3\,probes) = \left(\frac{n}{size}\right) \cdot \left(\frac{n-1}{size-1}\right) \cdot \left(\frac{n-2}{size-2}\right).$$

On and on it goes. We extrapolate out to x probes:

$$P(x\,probes) = \left(\frac{n}{size}\right) \cdot \left(\frac{n-1}{size-1}\right) \cdot \ldots \cdot \left(\frac{n-x+1}{size-x+1}\right).$$

This sequence would be smaller than assuming a probability of L for every missed probe. We could express this relationship with the following equation:

$$P(x\,probes) = \left(\frac{n}{size}\right) \cdot \left(\frac{n-1}{size-1}\right) \cdot \ldots \leq L \cdot L \ldots = L^x.$$

This gives us the probability of having multiple failed probes. We now want to think about the expected number of probes. One failed probe has the probability of L, and having more failed probes is less likely. To calculate the expected number of probes, we need to add the probabilities of all numbers of probes. So the P(1 probe) + P(2 probes)...on to infinity. You can think of this as a weighted average of all possible numbers of probes. A more likely number of probes contributes more to the weighted average. It turns out that we can calculate a value for this infinite series. The sum will converge to a specific value. We arrive at the following formula using the geometric series rule to give a bound for the number of probes in an unsuccessful search:

$$\text{Expected number of probes is less than } \sum_{x=0}^{\infty} L^x = \frac{1}{1-L}.$$

This equation bounds the expected number of probes or comparisons in an unsuccessful search. If 1/(1−L) is constant, then searches have an average case runtime complexity of O(1). We saw this in our analysis of linear probing where the performance was even worse than for the ideal uniform hashing.

For one final piece of analysis, look at the plot of 1/(1−L) between 0 and 1. This demonstrates just how critical the load factor can be in determining the expected complexity of hashing. This shows that as the load gets very high, the cost rapidly increases.

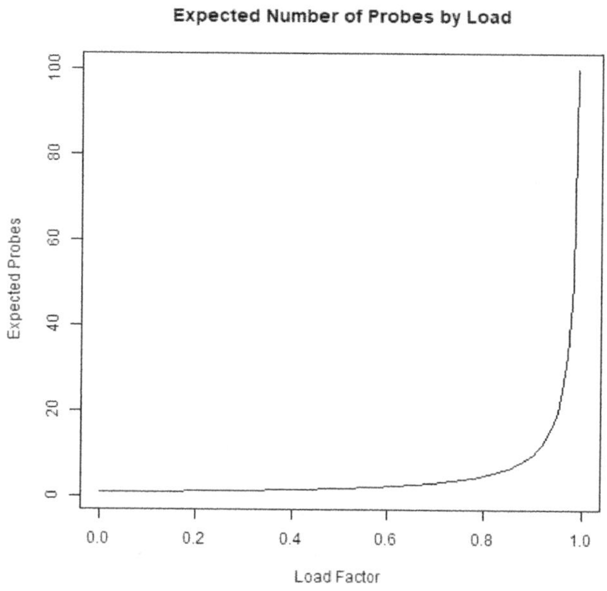

Figure 7.15

For completeness, we will present the much better performance of a successful search under uniform hashing:

$$\text{Expected number of probes} = \frac{1}{L}\ln\, ln\left(\frac{1}{1-L}\right).$$

Successful searches scale much better than unsuccessful ones, but they will still approach O(n) as the load gets high.

Chaining

An alternative strategy to open addressing is known as **chaining** or **separate chaining**. This strategy uses **separate linked lists** to handle collisions. The nodes in the linked list are said to be "chained" together like links on a chain. Our records are then organized by keeping them on "separate chains." This is the metaphor that gives the data structure its name. Rather than worrying about probing sequences, chaining will just keep a list of all records that collided at a hash index.

This approach is interesting because it represents an extremely powerful concept in data structures and algorithms, **composition**. Composition allows data structures to be combined in multiple powerful ways. How does it work? Well, data structures hold data, right? What if that "data" was another data structure? The specific composition used by separate chaining is an **array of linked lists**. To better understand this concept, we will visualize it and work through an example. The following image shows a chaining-based hash table after 3 add operations. No collisions have occurred yet:

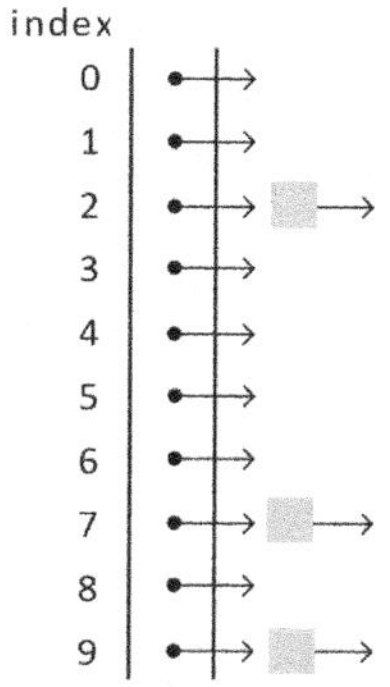

Figure 7.16

The beauty of separate chaining is that both adding and removing records in the table are made extremely easy. The complexity of the add and remove operations is delegated to the linked list. Let's assume the linked list supports add and remove by key operations on the list. The following functions give example implementations of add and remove for separate chaining. We will use the same Student class and the simple hash function that returns the key mod size.

The add function is below. Keep in mind that table[index] here is a linked list object:

```
function add(student)
    set index to hash(student.key)
    table[index].add(student)
```

Here is the remove function that, again, relies on the linked list implementation of remove:

```
function remove(key)
    set index to hash(key)
    table[index].remove(key)
```

When a Student record needs to be added to the table, whether a collision occurs or not, the Student is simply added to the linked list. See the diagram below:

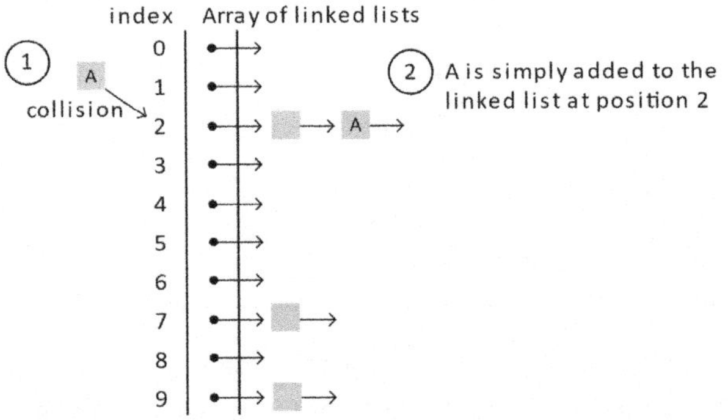

Figure 7.17

When considering the implementation, collisions are not explicitly considered. The hash index is calculated, and student A is inserted by asking the link list to insert it. Let's follow a few more add operations.

Suppose a student, B, is added with a hash index of 2.

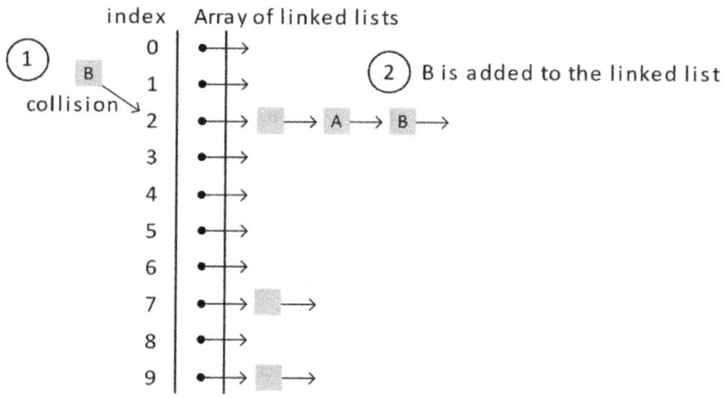

Figure 7.18

Now if C is added with a hash index of 3, it would be placed in the empty list at position 3 in the array.

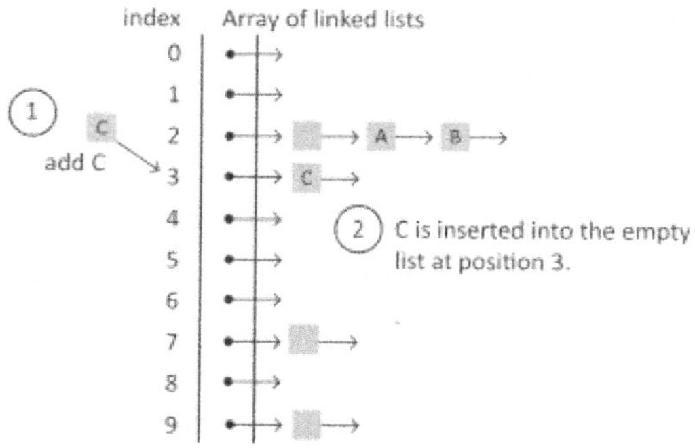

Figure 7.19

Here, the general idea of separate chaining is clear. Maybe it is also clear just how this could go wrong. In the case of search operations, finding the student with a given key would require searching for every student in the corresponding linked list. As you know from chapter 4, this is called **Linear Search**, and it requires O(n) operations, where n is the number of items in the list. For the separate chaining hash table, the length of any of those individual lists is hoped to be a small fraction of the total number of elements, n. If collisions are very common, then the size of an individual linked list in the data structure would get long and approach n in length. If this can be avoided and every list stays short, then searches on average take a constant number of operations leading to add, remove, and search operations that require O(1) operations on average. In the next section, we will expand on our implementation of a separate chaining hash table.

Separate Chaining Implementation

For our implementation of a separate chaining hash table, we will take an object-oriented approach. Let us assume that our data are the Student class defined before. Next, we will define a few classes that will help us create our hash table.

We will begin by defining our linked list. You may want to review chapter 4 before proceeding to better understand linked lists. We will first define our Node class and add a function to return the key associated with the student held at the node. The node class holds the connections in our list and acts as a container for the data we want to hold, the student data. In some languages, the next variable needs to be explicitly declared as a **reference or pointer** to the next Node.

```
class Node
    Student student
    Node next

    function key()
        return student.key
```

We will now define the data associated with our LinkedList class. The functions are a little more complex and will be covered next.

```
class LinkedList
    Node head
    Node tail
```

Our list will just keep track of references to the head and tail Nodes in the list. To start thinking about using this list, let's cover the add function for our LinkedList. We will add new students to the end of our list in constant time using the tail reference. We need to handle two cases for add. First, adding to an empty list means we need to set our head and tail variables correctly. All other cases will simply append to the tail and update the tail reference.

```
function add(student)
    # wrap student in a list node
    set newNode to an instance of Node
    set newNode.student to student

    if head equals null
        # empty list case
        set head to newNode
        set tail to newNode
    else
        # non-empty list case
        set tail.next to newNode
        set tail to newNode
```

Searching in the list will use **Linear Search**. Using the currentNode reference, we check every node for the key we are looking for. This will give us either the correct node or a null reference (reaching the end of the list without finding it).

```
function search(key)
    set currentNode to head
    while currentNode is not null and currentNode.key() is not key
        set currentNode to currentNode.next

    return currentNode
```

You may notice that we return currentNode regardless of whether the key matches or not. What we really want is either a Student object or nothing. We sidestepped this problem with open addressing by returning −1 when the search failed or the index of the student record when it was found. This means upstream code needs to check for the −1 before doing something with the result. In a similar way here, we send the problem upstream. Users of the code will need to check if the returned node reference is null. There are more elegant ways to solve this problem, but they are outside of the scope of the textbook. Visit the Wikipedia article on the Option Type for some background. For now, we will ask the user of the class to check the returned Node for the Student data.

To finish our LinkedList implementation for chaining, we will define our remove function. As remove makes modifications to our list structure, we will take special care to consider the different cases that change the head and tail members of the list. We will also use the convention of returning the removed node. This will allow the user of the code to optionally free its memory.

```
function remove(key)
    set previousNode to head
    set currentNode to head

    # traverse list, keep track of the previous node
    while currentNode is not null and currentNode.key() is not key
        set previousNode to currentNode
        set currentNode to currentNode.next

    # the key is found
    if currentNode is not null and currentNode.key() equals key
        # remove head
        if currentNode equals head
            set head to currentNode.next
            return currentNode
        # remove tail
        else if currentNode equals tail
            set tail to previousNode
            set previousNode.next to null
            return currentNode
        # removed interior node
        else
            set previousNode.next to currentNode.next
            return currentNode

    # the key could not be found
    return null
```

Now we will define our hash table with separate chaining. In the next code piece, we will define the data of our HashTable implemented with separate chaining. The HashTable's main piece of data is the composed array of LinkedLists. Also, the simple hash function is defined (key mod size).

```
class HashTable
    LinkedList Array table
    Integer size

    function hash(key)
        return key mod size
```

Here the simplicity of the implementation shines. The essential operations of the HashTable are delegated to LinkedList, and we get a robust data structure without a ton of programming effort! The functions for the add, search, and remove operations are presented below for our chaining-based HashTable:

```
function add(student)
    set index to hash(student.key)
    table[index].add(student)
```

```
function search(key)
    set index to hash(key)
    set studentNode to table[index].search(key)
    if studentNode is not null
        return studentNode.student
    else
        print "Student not found"
```

One version of remove is provided below:

```
function remove(key)
    set index to hash(key)
    set removeNode to table[index].remove(key)
    # optionally, free memory
    if removeNode is not null
        free removeNode
```

Some implementations of remove may expect a node reference to be given. If this is the case, remove could be implemented in constant time assuming the list is doubly linked. This would allow the node to be removed by immediately accessing the next and previous nodes. We have taken the approach of using a singly linked list and essentially duplicating the search functionality inside the LinkedList's remove function.

Not bad for less than 30 lines of code! Of course, there is more code in each of the components. This highlights the benefit of composition. Composing data structures opens a new world of interesting and useful data structure combinations.

Separate Chaining Complexity

Like with open addressing methods, the worst-case performance of search (and our remove function) is O(n). Probing would eventually consider nearly all records in our HashTable. This makes the O(n) complexity clear. Thinking about the worst-case performance for chaining may be a little different. What would happen if all our records were hashed to the same list? Suppose we inserted n Students into our table and that they all shared the same hash index. This means all Students would be inserted into the same LinkedList. Now the complexity of these operations would all require examining nearly all student records. The complexity of these operations in the HashTable would match the complexity of the LinkedList, O(n).

Now we will consider the average or expected runtime complexity. With the assumption that our keys are hashed into indexes following a simple uniform distribution, the hash function should, on average, "evenly" distribute the records along all the lists in our array. Here "evenly" means approximately evenly and not deviating too far from an even split.

Let's put this in more concrete terms. We will assume that the array for our table has size positions, and we are trying to insert n elements into the table. We will use the same load factor L to represent the load of our table, L = n / size. One difference from our open addressing methods is that now our L variable could be greater than 1. Using linked lists means that we can store more records in all of the lists than we have positions in our array of lists. When n Student records have been added to our chaining-based HashTable, they should be approximately evenly distributed between all the size lists in our array. This means that the n records are evenly split between size positions. On average, each list contains approximately L = n / size nodes. Searching those lists would require O(L) operations. The expected runtime cost for an unsuccessful search using chaining is often represented as O(1 + L). Several textbooks report the complexity this way to highlight the fact that when L is small (less than 1) the cost of computing the initial hash dominates the 1 part of the O(1 + L). If L is large, then it could dominate the complexity analysis. For example, using an array of size 1 would lead to L = n / 1 = n. So we get O(1 + L) = O(1 + n) = O(n). In practice, the value of L can be kept low at a small constant. This makes the average runtime of search O(1 + L) = O(1 + c) for some small constant c. This gives us our average runtime of O(1) for search, just as we wanted!

For the add operation, using the tail reference to insert records into the individual lists gives O(1) time cost. This means adding is efficient. Some textbooks report the complexity of remove or delete to be O(1) using a doubly linked list. If the Node's reference is passed to the remove function using this implementation, this would give us an O(1) remove operation. This assumes one thing though. You need to get the Node from somewhere. Where might we get this Node? Well, chances are that we get it from a search operation. This would mean that to effectively remove a student by its key requires O(1 + L) + O(1) operations. This matches the performance of our implementation that we provided in the code above.

The space complexity for separate chaining should be easy to understand. For the number of records we need to store, we will need that much space. We would also need some extra memory for references or pointer variables stored in the nodes of the LinkedLists. These "linking" variables increase overall memory consumption. Depending on the implementation, each node may need 1 or 2 link pointers. This memory would only increase the memory cost by a constant factor. The space required to store the elements of a separate chaining HashTable is O(n).

Design Trade-Offs for Hash Tables

So what's the catch? Hash tables are an amazing data structure that has attracted interest from computer scientists for decades. These hashing-based methods have given a lot of benefits to the field of computer science, from variable lookups in interpreters and compilers to fast implementations of sets, to name a few uses. With hash tables, we have smashed the already great search performance of Binary Search at O(log n) down to the excellent average case performance of O(1). Does it sound too good to be true? Well, as always, the answer is "It depends." Learning to consider the performance trade-offs of different data structures and algorithms is an essential skill for professional programmers. Let's consider what we are giving up in getting these performance gains.

The great performance scaling behavior of search is only in the average case. In practice, this represents most of the operations of the hash tables, but the possibility for extremely poor performance exists. While searching on average takes O(1), the worst-case time complexity is O(n) for all the methods we discussed. With open addressing methods, we try to avoid O(n) performance by being careful about our load factor L. This means that if L gets too large, we need to remove all our records and re-add them into a new larger array. This leads to another problem, wasted space. To keep our L at a nice value of, say, 0.75, that means that 25% of our array space goes unused. This may not be a big problem, but that depends on your application and system constraints. On your laptop, a few missing megabytes may go unnoticed. On a satellite or embedded device, lost memory may mean that your costs go through the roof. Chaining-based hash tables do not suffer from wasted memory, but as their load factor gets large, average search performance can suffer also. Again, a common practice is to remove and re-add the table records to a larger array once the load crosses a threshold. It should be noted again though that separate chaining already requires a substantial amount of extra memory to support linking references. In some ways, these memory concerns with hash tables are an example of the **speed-memory trade-off**, a classic concept in computer science. You will often find that in many cases you can trade time for space and space for time. In the case of hash tables, we sacrifice a little extra space to speed up our searches.

Another trade-off we are making may not be obvious. Hash tables guarantee only that searches will be efficient. If the order of the keys is important, we must look to another data structure. This means that finding the record with key 15 tells us nothing about the location of records with key 14 or 16. Let's look at an example to better understand this trade-off in which querying a range might be a problem for hash tables compared to a Binary Search.

Suppose we gave every student a numerical identifier when they enrolled in school. The first student got the number 1, the second student got the number 2, and so on. We could get every student that enrolled in a specific time period by selecting a range. Suppose we used chaining to store our 2,000 students using the identifier as the key. Our 2,000 students would be stored in an array of lists, and the array's size is 600. This means that on average each list contains between 3 and 4 nodes (3.3333...). Now we need to select 20 students that were enrolled at the same time. We need all the records for students whose keys are between 126 to 145 (inclusive). For a hash table, we would first search for key 126, add it to the list, then 127, then 128, and so on. Each search takes about three operations, so we get approximately 3.3333 * 20 = 66.6666 operations. What would this look like for a Binary Search? In Binary Search, the array of records is already sorted. This means that once we find the record with key 126, the record with key 127 is right next to it. The cost here would be $\log_2(2000) + 20$. This supposes that we use one Binary Search and 20 operations to add the records to our return list. This gives us approximately $\log_2(2000) + 20 = 10.9657 + 20 = 30.9657$. That is better than double our hash table implementation. However, we also see that individual searches using the hash table are over 3 times as fast as the Binary Search (10.6657 / 3.333 = 3.2897).

Exercises

1. On a sheet of paper, draw the steps of executing the following set of operations on a hash table implemented with open addressing and probing. Draw the

table, and make modifications after each operation to better understand clustering. Keep a second table for the status code.

a. Using **linear probing** with a table of size 13, make the following changes: add key 12; add key 13; add key 26; add key 6; add key 14, remove 26, add 39.

b. Using **quadratic probing** with a table of size 13, make the following changes: add key 12; add key 13; add key 26; add key 6; add key 14, remove 26, add 39.

c. Using **double hashing** with a table of size 13, make the following changes: add key 12; add key 13; add key 26; add key 6; add key 14, remove 26, add 39.

2.

Implement a hash table using linear probing as described in the chapter using your language of choice, but substitute the Student class for an integer type. Also, implement a utility function to print a representation of your table and the status associated with each open slot. Once your implementation is complete, execute the sequence of operations described in exercise 1, and print the table. Do your results match the paper results from exercise 1?

3.

Extend your linear probing hash table to have a load variable. Every time a record is added or removed, recalculate the load based on the size and the number of records. Add a procedure to create a new array that has size*2 as its new size, and add all the records to the new table. Recalculate the load variable when this procedure is called. Have your table call this rehash procedure anytime the load is greater than 0.75.

4. Think about your design for linear probing. Modify your design such that a quadratic probing HashTable or a double hashing HashTable could be created by simply inheriting from the linear probing table and overriding one or two functions.

5. Implement a separate chaining-based HashTable that stores integers as the key and the data. Compare the performance of the chaining-based hash table with linear probing. Generate 100 random keys in the range of 1 to 20,000, and add them to a linear probing-based HashTable with a size of 200. Add the same keys to a chaining-based HashTable with a size of 50. Once the tables are populated, time the execution of conducting 200 searches for randomly generated keys in the range. Which gave the better performance? Conduct this test several times. Do you see the same results? What factors contributed to these results?

References

Cormen, Thomas H., Charles E. Leiserson, Ronald L. Rivest, and Clifford Stein. *Introduction to Algorithms*, 2nd ed. Cambridge, MA: The MIT Press, 2001.

Dumey, Arnold I. "Indexing for Rapid Random Access Memory Systems," *Computers and Automation* 5, no. 12 (1956): 6–9.

Flajolet, P., P. Poblete, and A. Viola. "On the Analysis of Linear Probing Hashing," *Algorithmica* 22, no. 4 (1998): 490–515.

Knuth, Donald E. "Notes on 'Open' Addressing." 1963. https://jeffe.cs.illinois.edu/teaching/datastructures/2011/notes/knuth-OALP.pdf.

Malik, D. S. *Data Structures Using C++*. Cengage Learning, 2009.

8. Search Trees

Learning Objectives

After reading this chapter you will...

- extend your understanding of linked data structures.
- learn the basics techniques that drive performance in modern databases.

Introduction

We begin by listing some desirable aspects of data structures:

- economical and dynamic memory consumption
- ability to insert or delete keys in sublinear time
- ability to look up keys by exact match to a key in sublinear time
- ability to retrieve several key values based on a range

With these criteria in mind, let us review some of the data structures studied so far. Arrays allow us constant-time lookup, assuming we know the index of the item we want to retrieve. Finding that index requires a linear traversal of an unsorted array or a logarithmic Binary Search of a sorted array. Arrays have an unfortunate side effect in that they must be preallocated in memory.

As a result, inserts and deletes require either excess allocation or reallocation (with a great deal of copying).

We then considered linked lists. Conveniently, we are not required to know the capacity before performing the initial insert. Linked lists are less economic in memory consumption, as each stored datum requires us to also store a next (and possibly previous) reference. Locating a particular position for insert or delete requires a Linear Search, but the insertions or deletions at that point are constant time.

Then we arrived at hash tables. Finally, we had a data structure that allowed for true constant-time lookups based on a key. Inserts and deletes were also constant-time operations, assuming a sufficient hash function. These improvements were substantial but left us no option for retrieval of values based on a range.

What we want is some general-purpose data structure that maximizes the desired utility of linked lists while minimizing the rigidity of arrays and hash tables. Binary search trees fit nicely into this niche. Reusing some concepts we have learned so far, we can achieve sublinear times for inserts, deletions, and retrievals. We can grow and shrink our size as needed. We will require more storage than arrays but will not require the excess capacity as with hash tables.

Brief Introduction to Trees

Binary search trees are a subclass of binary trees, which are a subclass of trees, which are a subclass of graphs. Here we will only introduce enough details to facilitate an understanding of binary search trees. In chapter 11, we will provide more precise mathematical definitions of graphs and trees.

Trees (as well as graphs in general) consist of nodes and edges. As a note, nodes are also referred to as vertices (or vertex in the singular form). We will use nodes as containers for data, such as an integer, string, or even a database record. Nodes are related to other nodes via edges. Each edge connects two nodes and describes the relationship between those nodes. Edges in binary trees are child/parent relationships. One node is the parent, and the other is a child. Each node has at most one parent. A tree will have exactly one node without a parent. This node is called the root. Each node has no more than two children. A node with zero children is called a leaf. From time to time, we may consider a subtree, which is any given node and all its descendants. Although this chapter will focus mainly on binary trees, you should note the term "m-ary tree," where m is any positive integer and represents the maximum number of children for any given node.

Now we consider some additional tree-related terms. It is important to note that these terms are not consistently defined across different textbooks. In order to be consistent with another source you may likely read (Wikipedia), I will defer to the definitions found there. Whenever reading a new text, ensure that you first review that source's definition of terms:

- height—the number of nodes from a leaf node to the root, starting at 1
- depth—the number of nodes from the root to a particular node, starting at 1
- level—all descendants of the root that have the same depth
- full—a given m-ary tree is full if each node has exactly 0 or m children
- complete—a given m-ary tree is complete if every level is filled except possibly the last (which is filled from left to right)
- perfect—a given m-ary tree is perfect if it is full and all leaf nodes are at the same depth

Below is an example of a tree. Interior nodes are gray, and leaf nodes are white. The root node has been marked with an "R." Note that this is a ternary tree because any given node has at most three children. It is not full, which implies that it is neither complete nor perfect.

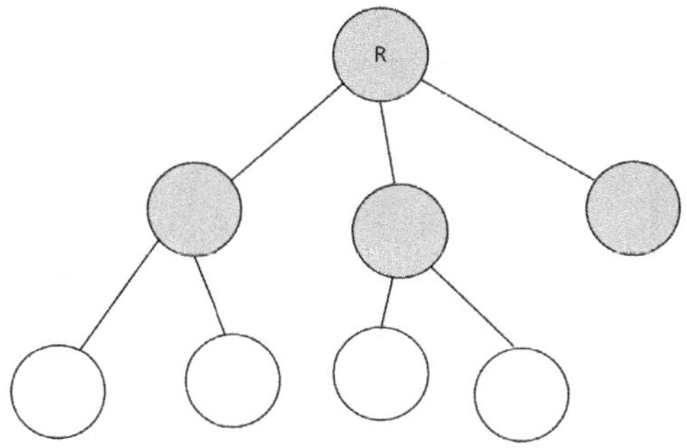

Figure 8.1

We may now make some useful assertions regarding binary trees:

- Because a perfect binary search tree implies that every interior node has two children, the number of nodes (n) is $2^k - 1$, where k is the number of levels in the tree. In a related manner, the number of levels in a tree is the floor of $\log_2 n$.
- Of all nodes in a perfect binary search tree, roughly half are leaf nodes, and the other half are interior. Precisely, the number of leaf nodes will be the ceiling of n/2, and the number of interior nodes will be the floor of n/2.

So far this concept of a tree does not produce much benefit. We could assign a key to each node, but what exactly would that mean? What does the relationship between parent and child imply? To derive value from trees, binary trees are insufficient, and we must apply more constraints.

A **binary search tree** (BST) is a specific type of binary tree that ensures that

- each node (N) is assigned a key.
- each node has a left child (L), which represents the subtree rooted at node L. The key of every node in this subtree is less than the key stored in node N. It is possible that a given node has no left child.
- each node has a right child (R), which represents the subtree rooted at R. The key of every node in this subtree is greater than the key in node N. It is possible that a given node has no right child.

Figure 8.2 is an example of a BST. The key stored in each node is an integer but could be of any data type that can be sorted. For convenience, we are assuming that BSTs do not contain duplicate keys, although we do not exactly need to. The tree below is perfect, but a BST does not need to be. As we discuss BSTs further, we will start to consider more problematic configurations.

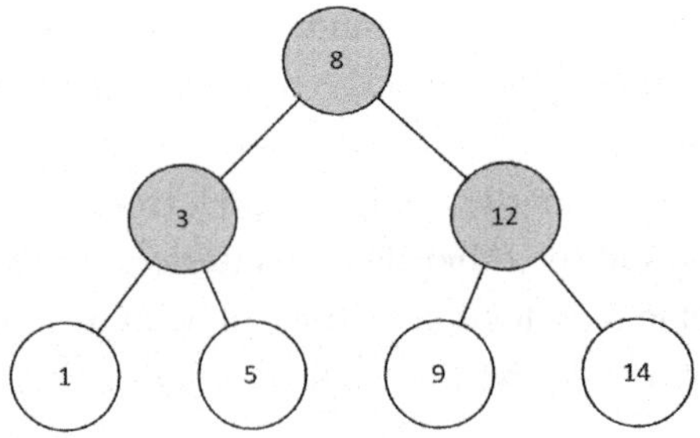

Figure 8.2

To understand the structure of a BST better, we can

consider an in-order traversal of the tree. This traversal is one in which we recursively visit the left child, current node, and right child. If you were to print each key during an in-order traversal, the result would be all keys from the tree in ascending order.

As implied in the figure above, nodes are modeled using the following class. In most practical applications, the Key property would hold some data other than type integer. Regardless, the simplicity of this model will be useful for the remainder of the chapter.

```
class Node
    Integer Key
    Node LeftChild
    Node RightChild
```

Searching

Searching is a simple operation. You begin at the root node considering a key (x) you want to find. If the key stored at the node is x, you have found it. If x is less than the node's key, you search the left subtree. If x is greater than the node's key, you search the right subtree. You continue this process until you arrive at a leaf node and have no more children to consider. Below is the pseudocode to further clarify the algorithm. The function is originally called with the root node, which then changes to descendants in the recursive calls.

```
function search(value, node)
    if value equals node.Key
        return FOUND
    else if value < node.Key and node has LeftChild
        return search(value, node.LeftChild)
    else if value > node.Key and node has RightChild
        return search(value, node.RightChild)
    else
        return NOTFOUND
```

Next, we should consider the runtime of searching a BST.

Just as with searching arrays and linked lists, we want to consider the amount of work necessary as the size of the data structure increases. In our recursive example above, we perform between 1 and 5 comparisons on each call to search (depending on exactly how you count). As a result, we have no more than 5 comparisons for each node visited. Because 5 is not dependent on the overall number of nodes, the amount of work to perform for each node visited is constant with respect to n.

How many nodes must we visit in the worst-case scenario? If we compare the desired value to the key at the root node and do not find the value, we have immediately eliminated roughly half of the values in our tree. Once at the second level, we perform the comparison again and eliminate half of this subtree, which was in turn half of the original. As a result, we reduce the number of keys we have to consider by half each time we visit a child. At worst, we will have to visit only one node in each level of the tree, resulting in ceiling $\log_2 n$ nodes visited. With O(log n) nodes visited and O(1) amount of work at each node, search can be run in O(log n) time for a perfect BST.

Insertion

To this point, we have assumed that a BST exists. We have yet to create one. Insertion simply searches for a valid position where the key would be if it existed and adds it at that position. In other words, the search for a nonexistent key always terminates in a leaf node. A naïve insertion algorithm considers this leaf node. If the key to insert is less than the leaf's key, you insert a new node as the left child. If the new key is greater than the leaf's key, you insert it as the right child. Pseudocode is included below to clarify:

```
function insert(value, node)
    if value equals node.Key
        # we have a duplicate key
        return FOUND
    else if value < node.Key
        if node has left child
            insert(value, node.LeftChild)
        else
            set node.LeftChild to new node with Key as value
    else if value > node.Key
        if node has right child
            insert(value, node.RightChild)
        else
            set node.RightChild to new node with Key as value
```

Deletion

When possible, it is good practice to enumerate the possible states that an algorithm may have to consider. When deleting a key from a BST, the node containing that key may be in four different states with respect to its children: no children, left child only, right child only, and both left and right children.

No Children

First, we must locate the node containing the key to be deleted. If that node has no children, it is by definition a leaf node. To delete a key at the leaf node, it suffices to simply remove the parent's reference to the leaf node. In the example below, the node containing the value 4 has no children. We can simply go to that node's parent and set the left child reference to null.

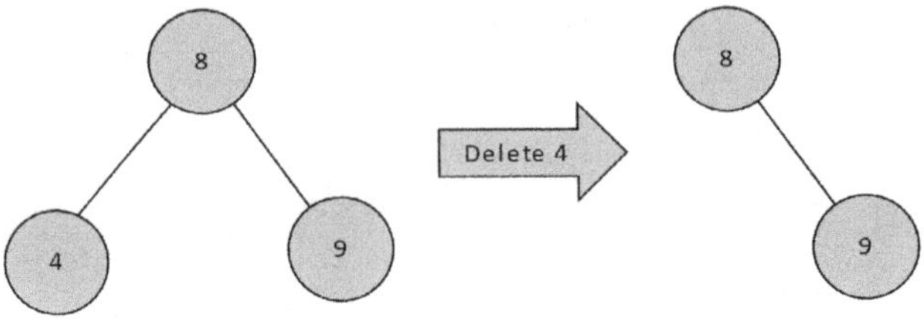

Figure 8.3

Left Child Only / Right Child Only

If a node contains a key to be deleted and has only one child, we can shift the appropriate subtree up. We can do this because all descendants in a node's left subtree are less than that node's value. In the case below, all left-side descendants of 8 are nodes containing values less than 8. If 4 only has one child, we can simply promote that child by setting 8's left child to that node. Similar reasoning would apply if 4 only had a right child.

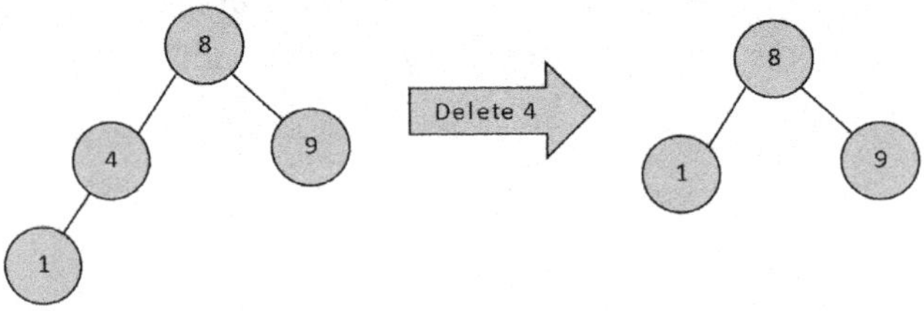

Figure 8.4

Both Left and Right Children

If a node containing a value to be deleted has both left and right children, we now must consider the possibility that those children may also be parents. This notably complicates the decision of what should be node 8's left child. If we were to shift the subtree starting at 1 up to 8's left child, that new node would have two right children (2 and 6), which obviously does not work. You would encounter a similar issue trying to promote 6 to be 8's left child. What we need instead is to find 4's in-order predecessor or in-order successor, remove that value from the tree (it is a leaf), and place it where 4 was. In the example below, we promoted 4's in-order predecessor, but we could have just as easily promoted the value 5.

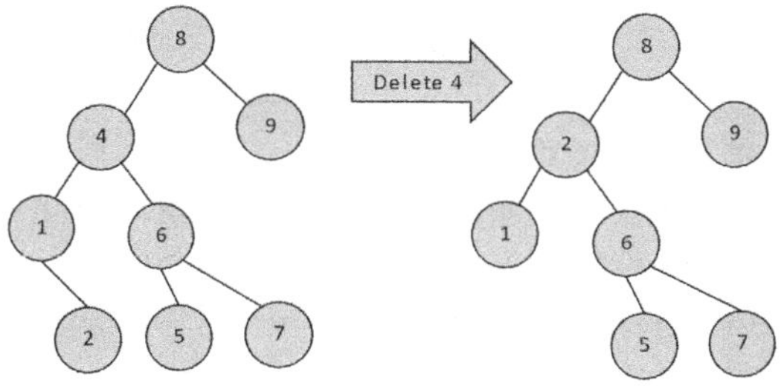

Figure 8.5

Unbalanced BSTs

When assessing the performance of search in BSTs, we had silently assumed that trees are perfect (or at least complete). We relied on this convenient property that the height of the tree was related to the logarithm of the number of nodes. In practice, this is rarely the case. Imagine we built a perfect BST using keys 1 through 7.

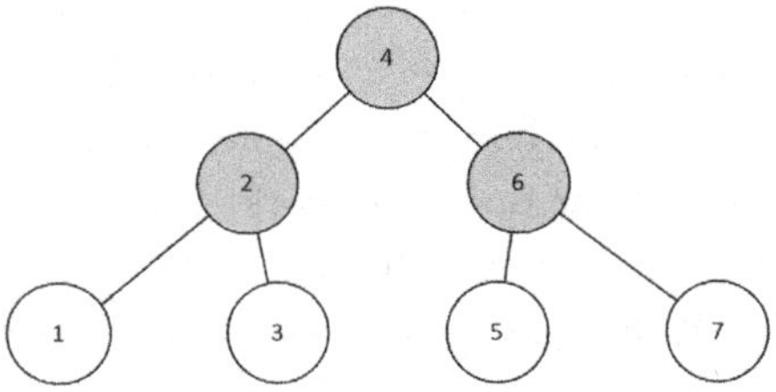

Figure 8.6

In figure 8.6, we can visualize the relationship between the number of nodes and the height of the tree. This relationship is logarithmic, so we can count on searching, inserting, and deleting keys to run in logarithmic time. However, if we perform inserts as specified above (in order from 1 to 7), we will actually end up with the tree in figure 8.7. Take a moment to trace the algorithm with pencil and paper to convince yourself this is the case.

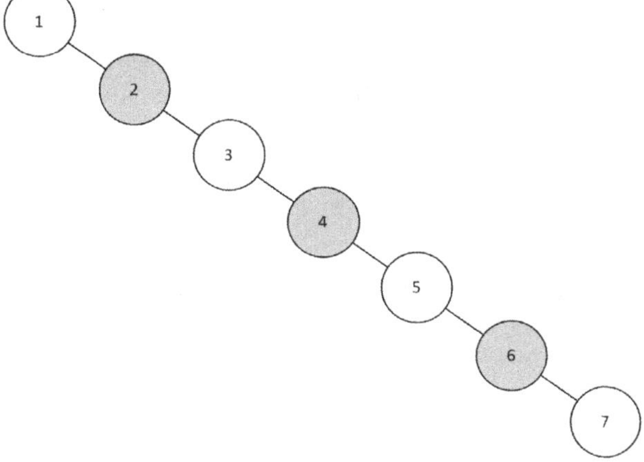

Figure 8.7

Our resulting structure, although technically a BST, also closely resembles a sorted linked list. When we studied linked lists, we were only able to search in linear time because all n nodes must be visited to ensure we found the desired key. The lesson learned is this: if we are not careful about how we perform inserts, we may likely construct a tree structure that cannot support O(log n) searches.

Ideally, we would love a tree to be perfect after each insert. This is not mathematically possible. If you have a perfect tree with seven nodes and three levels, inserting an eighth node will create a new level and result in a state where not all leaf nodes have the same depth. It may be desirable to maintain a complete tree. However, recall that complete trees must fill the lowest level from left to right. This constraint is not necessary, as we will be just as happy to fill it out right to left or in completely arbitrary order. As we can see, we need a new term to describe BSTs that allow for O(log n) searches and avoid the linked list type of configuration.

In a manner of speaking, we want our tree to be balanced after each insert. At this time, we will loosely define balance to be the condition such that the subtree heights of left and right subtrees are *roughly* equal. That leads to our next question: Can we modify our insert such that (1) the tree can remain balanced after each insert and (2) inserts can still be performed in O(log n) time?

Self-Balancing Trees

Self-balancing trees are those that maintain a balanced structure after each insertion and deletion and thus maintain an O(log n) search time. A thorough survey of these data structures could constitute chapters of text. This section will introduce how AVL trees maintain a balanced structure during insertion. We focus only on the insertion, but deletion must be addressed as well. Search, however, does not change from the naïve BST. Additional resources at the end of the chapter provide more information about this and other self-balancing trees.

AVL Trees

AVL trees are named after the computer scientists who developed them (G. M. Adelson-Velsky and E. M. Landis). After insertions that leave the tree in an unbalanced state, we achieve balance by performing small constant-timed adjustments called rotations.

First, we must determine whether an insertion has resulted in an unbalanced tree. To determine this, we use a metric called the balance factor. This integer is the difference in the heights of a node's left and right subtrees. Below is the simplest possible tree where we can witness such an imbalance. As usual, values are stored inside the node. Subtree heights are stored at the upper right of each node. If a left or right child does not exist, then the subtree height is 0. In this example, the node containing 8 has a left child with subtree height of 2 and a right child with subtree height of 0. The absolute difference between 2 and 0 is 2. This is above our threshold of 1, so our tree is unbalanced.

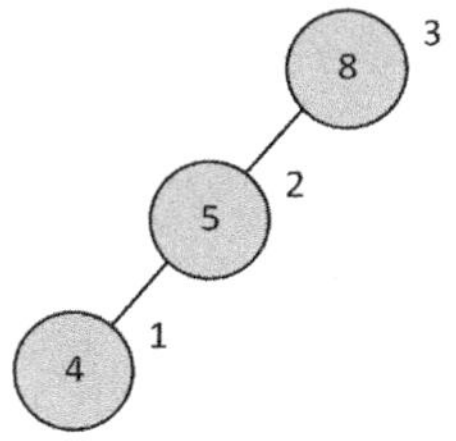

Figure 8.8

We now have a means for detecting unbalanced trees and are left to determine how to bring the tree back into balance. This is where we employ rotations. Rotations are small, constant-time adjustments to a subtree that improve the balance of that subtree. They are called rotations because they have the visual effect of rotating that subtree to a more balanced state. In figure 8.8, a rotation makes the 5 node the new root with a left child of 4 and a right child of 8. This is visually depicted in figure 8.9. Notice that after the rotation, the height of the subtree starting at 8 is now 1. Node 5 has left and right subtrees both at height 1. The difference is 0, which is not greater than 1, indicating that we are now in balance.

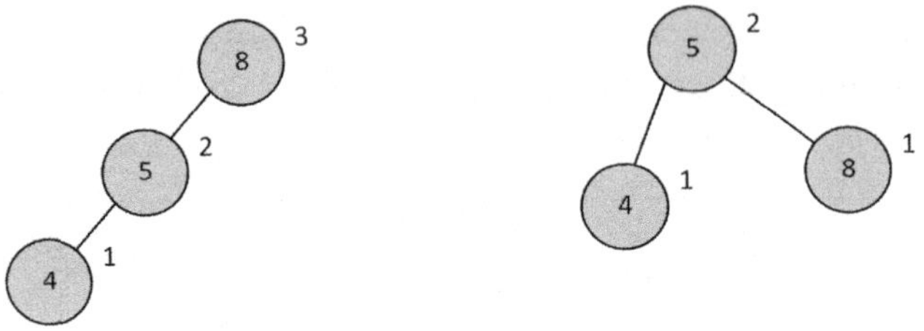

Figure 8.9

We make one last consideration regarding AVL trees. Earlier we had described this modification to BSTs as "self-balancing" and the heights of left and right subtrees as roughly equal. What actually happens is more nuanced and worthy of discussion. With perfect BSTs, we concluded that the relationship between the number of nodes in the tree and the number of comparisons required for a search was logarithmic. For AVL trees, we must be able to show the same relationship applies.

A proof by induction is able to show that the height of any AVL tree is O(log n). Note the distinction here. Perfect binary trees were shown to have a height equal to the ceiling of $\log_2$ n (or more precisely, $\log_2(n+1)$). AVL trees are said to have a height of O(log n), which is less precise. Rather than reviewing the inductive proof (which can easily be found online and in many reference textbooks), let us consider the following two trees:

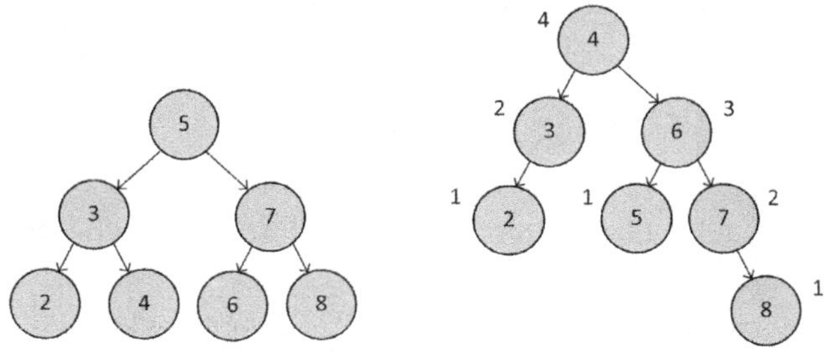

Figure 8.10

The tree on the left is a perfect BST. It has 7 nodes, which implies a height of $\log_2(7+1) = 3$. The tree on the right is a balanced AVL tree. Note that the height is no longer 3, even though we claim that the tree is balanced and that search times are still O(log n). We point this out to illustrate that while some algorithms may share the same Big-O classification, their actual runtimes may differ. Because of the rotations, we ensure that the difference in heights of the left and right subtrees is no more than 1. This then ensures that our AVL tree, while not complete or perfect, has a height no greater than $1 + \log_2(n+1)$. The additional 1 does not significantly impact the growth of the function as n becomes very large, so we can conclude that searching an AVL tree can still be accomplished in O(log n) time.

Exercises

1.

Drawing your own diagrams, perform insertions into an empty binary search tree. Can you determine the appropriate insertion sequence to produce

a. a perfect BST of size 7?
b. a BST where each node has only left children or no children?
c. a BST where each node has only right children or no children?

2.

In the language of your choice, implement BST deletes. Rather than solving the entire problem at once, break your code into three distinct cases:

a. Node to delete has no children.
b. Node to delete has one child.
c. Node to delete has two children.

References

Alexander, Eric. "AVL Trees." Computer Science User Pages. University of Wisconsin-Madison. Accessed September 27, 2023. https://pages.cs.wisc.edu/~ealexand/cs367/NOTES/AVL-Trees/index.html

"Binary Tree." Wikipedia. Last modified August 28 2023. https://en.wikipedia.org/wiki/Binary_tree

Galles, David. "AVL Tree." USF Computer Science Department. Accessed September 27, 2023. https://www.cs.usfca.edu/~galles/visualization/AVLtree.html

Galles, David. "Binary Search Tree." USF Computer Science Department. Accessed September 27, 2023. https://www.cs.usfca.edu/~galles/visualization/BST.html

9. Priority Queues

Learning Objectives

After reading this chapter you will…

- understand the concept of a priority queue.
- understand the features of heaps, one of the most common priority queue implementations.
- be able to implement Heap Sort, a sorting algorithm that uses a priority queue.
- be able to explain the common operations on priority queues and their complexity.
- be able to implement a binomial heap that supports a fast union operation.

Introduction

We have already discussed the concept of a queue. This is a data structure that accepts items and removes them in the order they were inserted. This is often referred to as first-in-first-out, or FIFO. A priority queue serves like a regular queue allowing items to be inserted, but it allows for the item with the highest priority to exit the queue first. We could implement a priority queue as a simple array with a current capacity that just resorts all the items by priority every time an item is inserted. This would mean that the insert operation for our simple priority queue would be O(n log n). There may be more clever approaches to preserve the sorted order by moving things over. This might lead to an O(n) insert operation. Extract could work similarly by removing the first element and then copying all the elements over. Could we do better than O(n) though? Linear time might be a long time to wait for a large n. For example, suppose many players are waiting to start an online game. We could use a priority queue to add players to the next game based on how long they have been waiting. A game AI may want to prioritize and target players that are the most dangerous first. We will see in chapter 11 that priority queues are used as the foundation for some important graph algorithms. Since this type of data structure could be very useful, researchers and engineers have discovered a variety of data structures that greatly improve the time-cost complexity. We will explore two interesting implementations of priority queues in this chapter.

Heaps

A heap is a data structure that guarantees that the minimum (or maximum) value is easily extracted. The most common heap is a binary heap, which is a sort of binary tree. In a binary heap, the "left" or "right" position of a child node no longer carries any specific meaning. Rather, in a max-binary-heap, or just max-heap, the parent is guaranteed to be greater than both children. Min-binary-heaps naturally reverse that relationship, with the parent guaranteed to be less than the children. We call this quality the heap property. It will allow us to isolate our reasoning to only subheaps and thus aid our understanding of heap-related algorithms.

For the remainder of this section, we assume max-binary-heaps to avoid confusion. We will also assume unique values in our binary heap. This simplifies the relationships between parents and their children. If a particular application of binary heaps necessitates duplicate keys, this is easily remedied by adjusting the appropriate comparisons. The figure below gives an example of a heap:

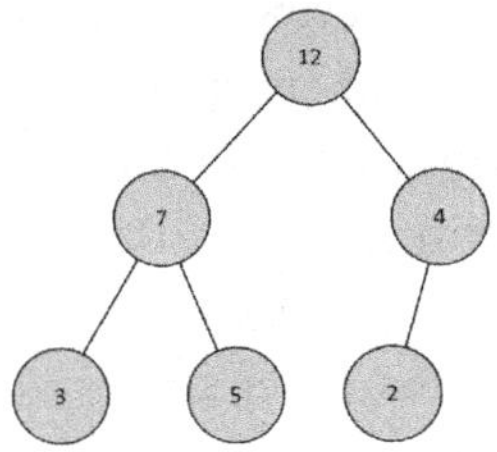

Figure 9.1

This distinction between parent and child nodes leads to two convenient properties of binary heaps:

- For a given heap, the maximum value (or key) is easily accessible at the root of the tree. As we will soon see, this implies not that it is easily extracted from the structure but simply that finding it is trivial.
- For any given node in a binary heap, all descendants contain values less than that node's value. In other words, any given subtree of a max-binary-heap is also a valid max-binary-heap.

Let us emphasize and further address a common point of confusion. Although binary heaps are binary tree structures, their similarities with binary *search* trees (BST) end there. Recall from chapter 8 that an in-order traversal of a binary search tree will produce a sorted result. This is true because, for any given node in a BST, all left descendants are less than that node, and all right descendants are greater than it. In binary heaps, the left descendant is less than the parent, and the right descendant is less than it as well, but there is no other defined relationship between the two descendants.

To understand insertion and extraction, first note the shape of the binary tree above. A tree is a complete binary tree if each node has two children and all levels are filled except possibly the last, which is filled from left to right (chapter 8). Using some clever tricks, we can store a complete binary tree as an array. Because each level of the tree is filled and the last is filled left to right, we can simply list all elements in level 0, followed by all elements in level 1, and so on. Once these values are stored in an array, some simple arithmetic on the indexes allows traversal from a node to its parents or its children. We will be regularly adding and removing data from the heap itself. As we have seen in prior chapters, arrays are an insufficient data structure for accomplishing this. For the sake of simplicity, we will assume excess capacity at the end of the array. In practice, we would probably use some sort of abstract list that is able to grow or shrink and provides constant-time lookups. This might be something like an array that automatically reallocates when its capacity is reached. Implementations of these lists exist in most modern languages. For the present discussion, we can just treat the underlying storage as a typical array. The image below shows a heap represented as an array with integer values for the priorities:

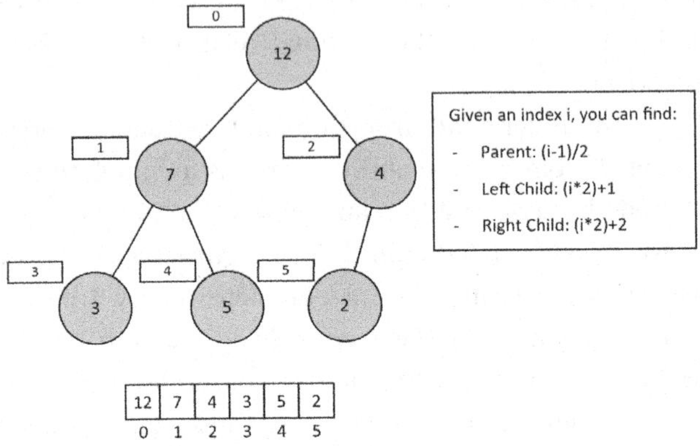

Figure 9.2

For now, we will work only with integers that serve as the priorities themselves. To extend this into a more useful data structure, we would need only to change the contents of the array to an object or object reference. This would allow us to hold a more useful structure such as a student record or a game player's data. Then the only other change needed would be to do comparisons on array[index].priority instead of just array[index]. This modification is like the one we discussed regarding Linear Search in chapter 4. Recognize that we can generalize this representation easily to accommodate data records that are slightly more sophisticated than just integers.

Operations on Binary Heaps

Before we discuss the general operations on binary heaps, let's discuss some helper functions that will help us with our array representation. We will define functions that will allow us to find the parent, left child, and right child indexes given the index for any node in the tree structure. These would be helpful to define for any tree structure that we wanted to implement using an array. For this representation, the root will always be at the index 0. These are given in the figure above, but we will provide them as code here:

```
function parent(index)
    return floor((index-1)/2)
```

Here the floor function is the same as the mathematical function floor. It rounds down to the next integer.

```
function leftChild(index)
    return index*2 + 1
```

```
function rightChild(index)
    return index*2 + 2
```

Heapify and Sift Up

Using the above functions to access positions in our tree, we can develop two important helper functions that will allow us to modify the tree and work to maintain our heap properties. These are the functions heapify and siftUp. When we are building our heap or modifying the priority of an item, these functions will be useful. The heapify function will exchange a parent with the larger of its children and then recursively heapify the subheap. The siftUp function will exchange a child with its parent to maintain the heap property by moving larger elements up the heap until it either is smaller than its parent or becomes the root node element. The siftUp function will be used when we want to increase the priority of an element. Let's look at the pseudocode for these functions in the context of an array-based max-heap implementation.

The heapify function below lets a potentially small value work its way down the max-heap to find its correct place in the heap ordering of the tree. This code uses a size parameter that gives the current number of elements in the heap. This code also makes use of an exchange function like the one discussed in chapter 3 on sorting. This simply switches the elements of an array using indexes.

```
function heapify(array, index, size)
    set leftIndex to leftChild(index)
    set rightIndex to rightChild(index)

    # check if the left child is larger than parent
    if leftIndex < size and array[leftIndex] > array[index]
        set largest to leftIndex
    else
        set largest to index

    # check if right is larger than current largest
    if rightIndex < size and array[rightIndex] > array[largest]
        set largest to rightIndex

    if largest is not index
        exchange(array, index, largest)
        # recursively heapify the next subheap
        heapify(array, largest, size)
```

Now we will present the siftUp function. This function works in the reverse direction from heapify. It allows for a node with a potentially large value to make its way up the heap to the correct position to preserve the heap property. Since siftUp moves items toward index 0, the size is not needed. This does assume that the given index is valid.

```
function siftUp(array, index)
    while index > 0 and array[parent(index)] < array[index]
        exchange(array, index, parent(index))
        set index to parent(index)
```

With these two helper functions, we can now implement the methods to insert elements and remove the max-element from our priority queue. Before we move on though, let's think about the complexity of these operations. Each of these methods moves items up or down the depths of a binary tree. If the tree could remain balanced, then a traversal from the top to bottom or bottom to top should only require O(log n) operations, assuming that the tree is balanced.

Insertion

Consider inserting the number 8 into the prior binary heap example. Imagine if we simply added that 8 in the array after the 2 (in position 6).

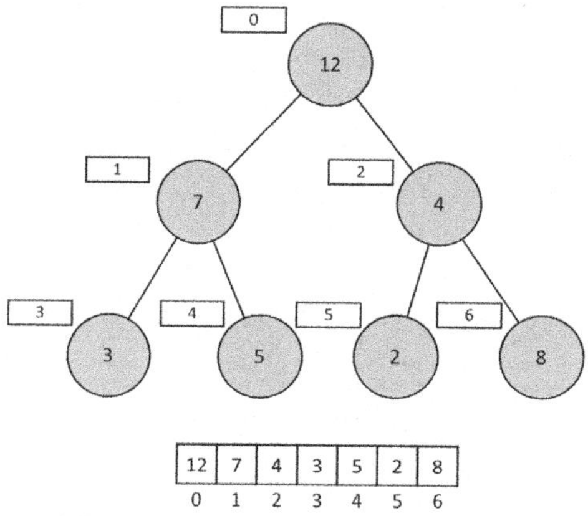

Figure 9.3

What can we now claim about the state of our binary heap? Subheaps starting at indexes 1, 3, 4, and 5 are all still valid subheaps because the heap property is preserved. In other words, our erroneous insertion of 8 under 4 does not alter the descendants of these 4 nodes. As a result, we could hypothetically leave these subheaps unaltered in our corrected heap. This leaves nodes at indexes 0 and 2. These are the two nodes that have had their descendants altered. The heap property between node indexes 2 and 6 no longer holds, so let us start there. If we were to switch the 4 and 8, we would restore the heap property between those two indexes. We also know that moving 8 into index 2 will not affect the heap property between indexes 2 and 5. If 2 was less than 4 and 4 was found to be less than 8, switching the 4 and 8 does not impact the heap property between indexes 2 and 5. Once the 4 and 8 are in the correct positions, we know that the subheap starting at index 2 is correct. From here, we simply perform the same operations on subsequent parents until the next parent's value is greater than the value we are trying to insert. Given that we are inserting 8 and our root node's value is 12, we can stop iterating at this point. The pseudocode below is descriptive but much simpler than practical implementations, which must consider precise data structures for storing the heap. This provides the general pattern for inserting into a binary heap regardless of the underlying implementation. We place the new element at the end of the heap and then essentially siftUp that element to the place that will preserve the heap property.

```
function insert(heap, value)
    Add value at the end of the heap
    set n to last element in heap
    while n is not heap.root and n.value > n.parent.value
        exchange n and n.parent within heap
        set n to n.parent
```

Runtime of insertion is independent of whether the heap is stored as object references or an array. In either case, we have to compare n.value to n.parent.value at most O(log n) times. Note that, unlike the caveat included in binary search trees (where traversing an unbalanced tree may be as slow as O(n)), binary heaps maintain their balance by building each new depth level before increasing its depth. This ensures that an imbalanced binary heap does not occur. This fact guarantees that our insert operation is O(log n).

A specific implementation for insert with our array-based heap is provided below:

```
function insert(array, value)
    # place the element at the end and update size
    set array[size] to value
    siftUp(array, size)
    set size to size + 1
```

Extraction

Extraction is the process of removing the root node of a binary heap. It works much the same way as insertion but in reverse. The general strategy is as follows.

To extract an element from the heap...

1. Extract the root element, and prepare to return it.
2. Replace the root with the last element in the heap.
3. Call heapify on the new root to correct any violations of the heap property.

As an example, consider our corrected heap from before. If we overwrite our 12 (at index 0) with the last value from the array (4 at index 6), the result will be as follows in figure 9.5. Just as we saw with insertion, many of our subheaps still have the heap property preserved. In fact, the only two places where the heap property no longer holds are from indexes 0 to 1 and 0 to 2. If the value at the root is less than the maximum value of its two children, then we swap the root value and that maximum. Wc will continue this process of pushing the root value down until the current node is greater than both children, thus conforming to the heap property. In this case, we swap the 4 with the 8. The root node conforms to the heap property because its children are 7 and 4. The node with value 4 (now at index 2) only has one child (2 at index 5). It conforms to the heap property, and our extraction is complete.

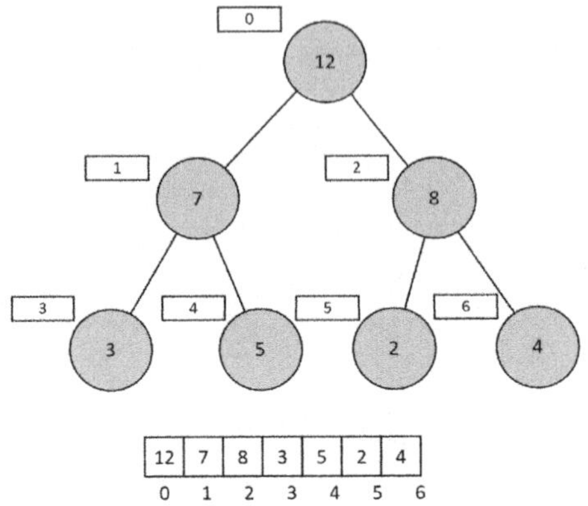

Figure 9.4

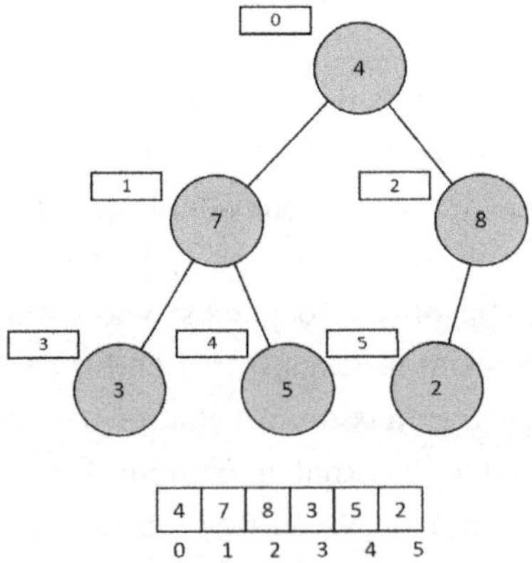

Figure 9.5

As before, the pseudocode is much simpler than the actual implementation.

```
function extract(heap)
    set v to heap.root.value
    set heap.root.value to heap.last.value
    Delete heap.last.value
    set n to heap.root
    while n has children and n.value < Max(n.left.value, n.right.value)
        if Max(n.left.value, n.right.value) equals n.left.value
            exchange n and n.left
            set n to n.left
        else
            exchange n and n.right
            set n to n.right
    return v
```

The array-based implementation could use the heapify function to give the following code:

```
function extract(array)
    # save the max item to return
    set maxItem to array[0]

    # move the last item to the root, and update size
    set array[0] to array[size - 1]
    set size to size - 1

    # now heapify the array from the root
    heapify(array, 0, size)

    return maxItem
```

While accessing the max-element would only require O(1) time, updating the heap after it is removed requires a call to heapify. This function requires O(log n). While not constant time, O(log n) is a great improvement over our initial naïve implementation ideas from the introduction. Our initial idea of sorting and then always copying moving elements up or down would have required O(n) operations to maintain our priority queue when inserting and removing elements. The max-heap greatly improves on these complexity estimates, giving O(log n) for both insert and extract.

Heap Sort

Heap Sort presents an interesting use of a priority queue. It can be used to sort the elements of an array. Once insertion and extraction have been defined, Heap Sort becomes a trivial step. We first build the heap, then repeatedly extract the maximum element and put it at the end of the array. Much like Selection Sort, Heap Sort will find the extreme value, place it into the correct position, then find the extreme of the remaining values. The trick is in how we perceive the heap. If we model it using an array, we can then sort the values in place by extracting the maximum. The extraction makes the heap smaller by one, but arrays are fixed size and still have the extra space allocated at the end. This portion at the end of the array becomes our sorted portion. As we perform more extractions and move those extracted values to the end of the array, our sorted portion gets bigger. See the figure below for an example. Unlike Selection Sort, where finding that extreme value requires an O(n) findMax or findMin, heaps allow us to extract the extreme value and revise our heap in O(log n) time. We perform this operation O(n) times, resulting in an O(n log n) sorting algorithm. The following figure gives an example execution of the sorting algorithm:

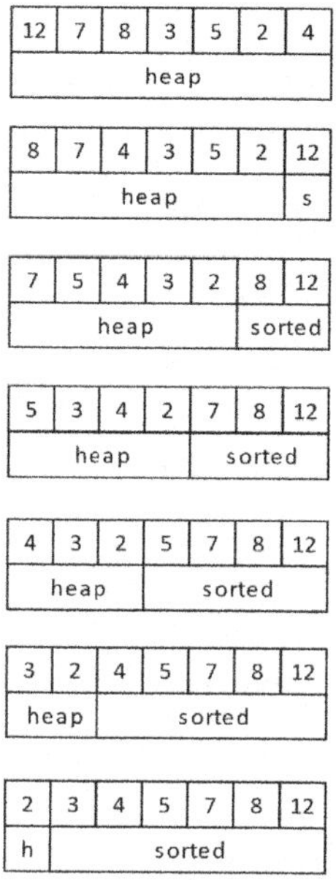

Figure 9.6

Before we give the implementation of Heap Sort, we should also mention how to build the heap in the first place. If we are given an array of random values, there is no guarantee that these will conform to our requirements of a heap. This is accomplished by calling the heapify function repeatedly to build valid heaps starting at the deeper levels of the balanced tree up to the root. Below is the code for buildHeap, which will take an array of elements to be sorted and put them into the correct heap ordering:

```
function buildHeap(array)
    set size to length(array)
    # move the index up to the root by counting down
    for index from size/2 to 0
        heapify(array, index, size)
```

It might be unintuitive, but buildHeap is O(n) in its time complexity. At first glance, we see heapify, an O(log n) operation, getting called inside a loop that runs from size/2 down to 0. This might seem like O(n log n). What we need to remember though is that O(log n) is a worst-case scenario for heapify. It may be more efficient. As we build the heap up, we start at position size/2. This is because half of the heap's elements will be leaves of the binary tree located at the deepest level. As we heapify the level just before the leaves, we only need to consider three elements: the parent and its two leaf children. As heapify runs, the amount of work is proportional to the height of the subtree it is operating on. Only on the very last call does heapify potentially visit all log n of the levels of the tree. We will omit the calculation details, but it has been proven that O(n) gives a tighter bound on the worst-case time complexity of buildHeap.

Now we are ready to implement Heap Sort. An implementation is provided below. Our O(n log n) complexity comes from calling heapify from the root every time we extract the next largest value. Another useful feature of this algorithm is that it is an **in-place** sorting algorithm. This means the extra space (auxiliary space) only consumes O(1) space in memory. So Heap Sort compares favorably to Quick Sort with a better worst-case complexity (O(n log n) vs. $O(n^2)$), and it offers an improvement over Merge Sort in terms of its auxiliary space usage (O(1) auxiliary space vs. O(n)). We should note that Heap Sort may perform poorly in practice due to cache misses, since traversing a tree skips around the elements of the array.

```
function heapSort(array)
    # heap order the elements
    buildHeap(array)
    # repeatedly extract elements and place them in position
    for index from (size - 1) to 1
        set element to extract(array)
        set array[index] to element
```

This section has provided an overview of heapSort and the

concept of a heap more generally. Heaps are great data structures for implementing priority queues. They can be implemented using arrays or linked data structures. The array implementation also demonstrates an interesting example of embedding a tree structure into a linear array. The power and simplicity of heaps make them a popular data structure. One potential disadvantage of the heap is that merging two heaps might require O(n) operation. To combine these two heaps, we would need to create a new array, recopy the elements, and then call buildHeap, taking O(n) operations. In the next section, we will discuss a new data structure that supports an O(log n) **union** operation.

Binomial Heaps

The **binomial heap** supports a fast union operation. When two heaps are given, union can combine them into a new heap containing all the combined elements from the two heaps. Binomial heaps are linked data structures, but they are a bit more complex than linked lists or binary trees. There is an interesting characteristic to their structure, which models the pattern of binary numbers, and combining them parallels binary addition. Binary numbers and powers of 2 seem to pop up everywhere in computer science. In this section, we will present the binomial heap and demonstrate how it can provide a fast union operation. Then we will see how many of the other operations on heaps can be implemented with clever use of the union function.

Linked Structures of Binomial Heaps

To build our heap, we need to discuss two main structures. The first part is the **binomial tree**. Each binomial tree is composed of connected **binomial nodes**. The structure of a binomial tree can be described recursively. Each binomial tree has a value k that represents its degree. The degree 0 tree, B_0, has one element and no children. A degree k tree, B_k, has k direct children but 2^k nodes in total. When the B_k tree is constructed, the roots of two B_{k-1} trees are examined. The largest root of the two trees is assigned as the root of the new tree, assuming a **max-binomial-heap**. The heap is then represented as a **list of binomial trees**. A collection of trees is known as a **forest**. The main idea of the binomial is that each heap is a list of trees, and to combine the two heaps, one just needs to combine all the trees of equal degree. To facilitate this, the trees are always ordered by increasing tree degrees. A few illustrations will help you understand this process a little better.

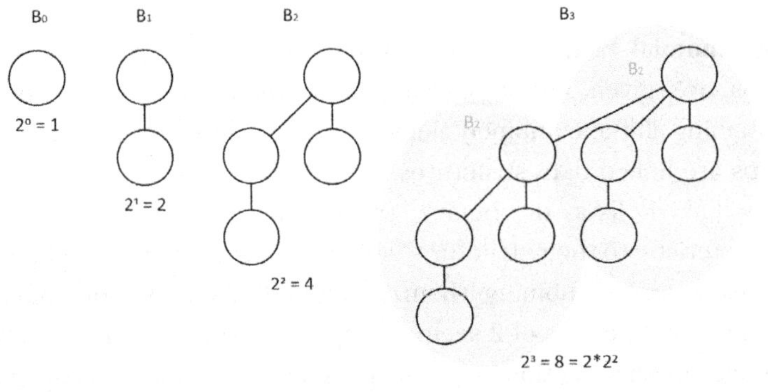

Figure 9.7

Using these trees, a heap would then be a list of these trees. To preserve the max-heap property, any node's priority must be larger than its child. Below is an example binomial heap. Let's call this $heap_A$:

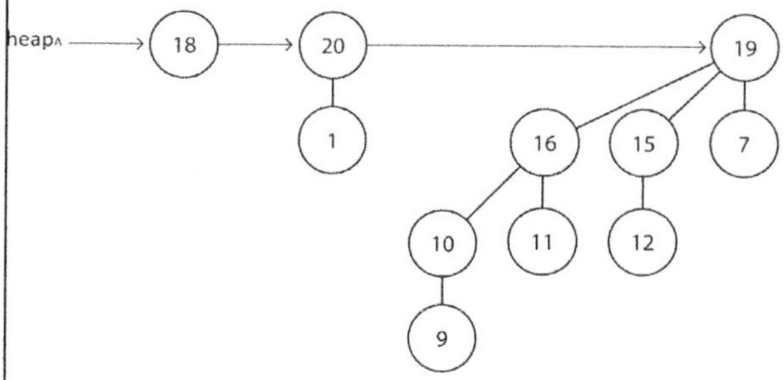

Figure 9.8

Notice that the maximum element is in B_1 tree of this heap. This illustrates that the actual max-heap element is one of the root nodes of trees in the list. Now we can make the connection to binary numbers. This heap will either have a tree of any given degree or not. This could be indicated by a 0 or 1. So the above heap has a degree 0 tree, a degree 1 tree, and a degree 3 tree. In binary with the bits correctly ordered, this would be 1011_2 or the number 11_{10} in base 10. Suppose there is another heap, $heap_B$, that we wish to merge with. This is given below:

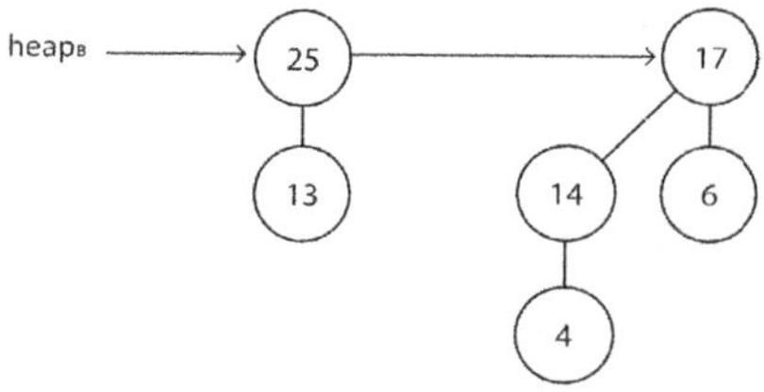

Figure 9.9

This heap has trees for degrees 1 and 2. In binary, we could represent this pattern as 110_2 or the number 6_{10} in base 10. We will soon look at how these heaps could be merged, but first let's consider the process of binary addition for these two binary numbers: 11 and 6. The figure below gives an example of the addition:

2^4	2^3	2^2	2^1	2^0
16	8	4	2	1

		1	1		
		1	0	1	1
+			1	1	0
	1	0	0	0	1

Figure 9.10

This diagram of binary addition also demonstrates how our trees need to be combined to create the correct structure for our unified trees. The following images will show these steps in action. First, we will merge the two lists into another list (but not a heap yet). This merge is similar to the merge operation in Merge Sort:

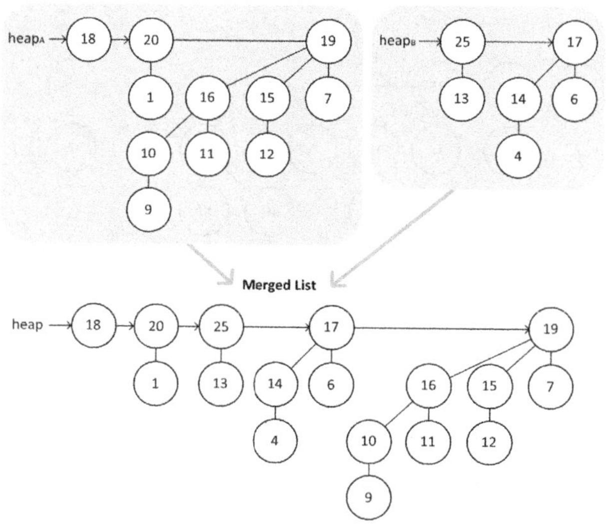

Figure 9.11

Next, the algorithm will examine two trees at a time to determine if the trees are the same degree. Any trees that have an equal degree will be merged. Because the nodes are ordered by degree, we only need to consider two nodes at a time and potentially keep track of a carry node.

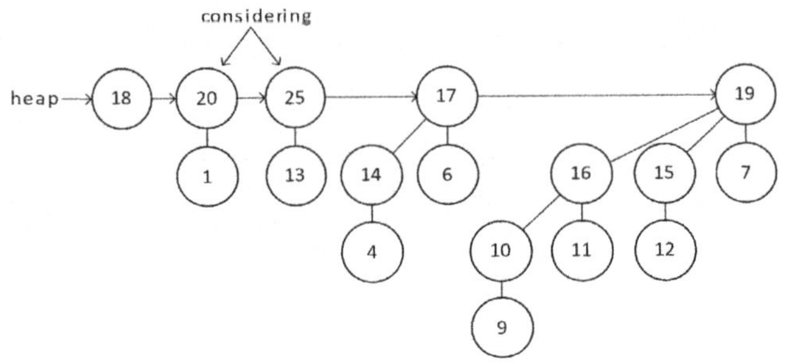

Figure 9.12

These nodes are not equal in degree. We can move on.

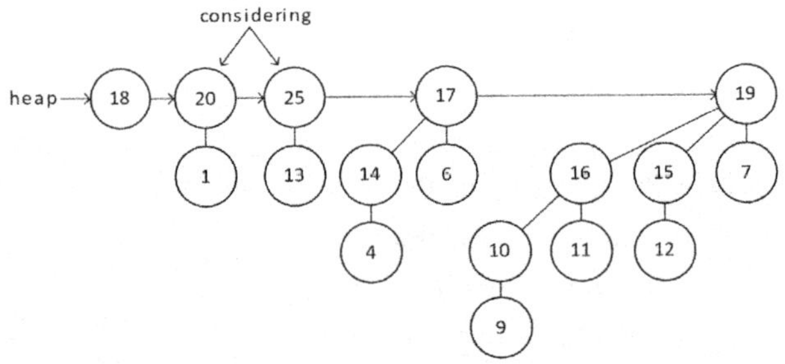

Figure 9.13

Now we are considering two nodes of equal degree. These two B_1 trees need to become a B_2 tree.

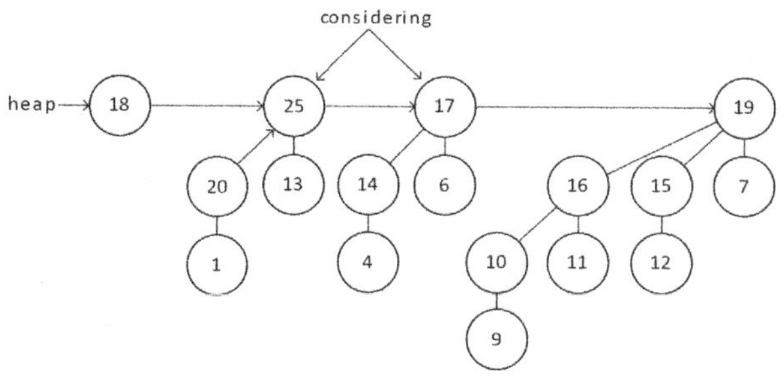

Figure 9.14

After combining the B_1 trees, we now have two B_2 trees that need to be combined. This will be done such that the maximum item of the roots becomes the new root.

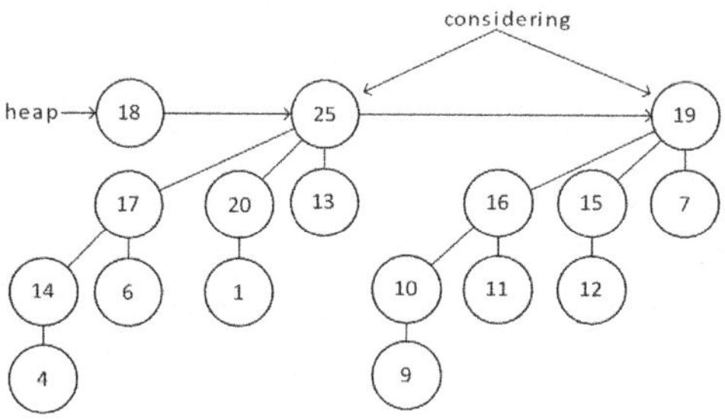

Figure 9.15

Again, the "carry" from addition means that we now have

two B3 trees to merge. This step creates a B4 tree with $2^4 = 16$ nodes. The final merged heap is below, with one B0 node and one B4 node:

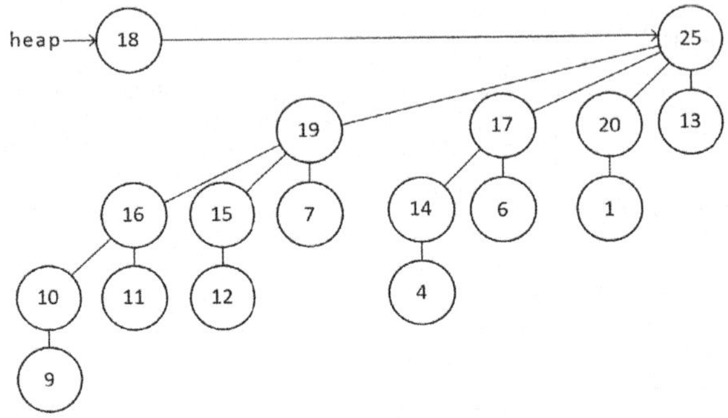

Figure 9.16

Implementing Binomial Trees

To implement Binomial trees, we will need to create a node class with appropriate links. Again, we will use references, also known as pointers, for our links. A node—and by extension, a tree—can be represented with the BinomialTree class below. In this class, Data would be any entity that needed storing in the priority queue:

```
Class BinomialTree
    Data data
    Integer priority
    Integer degree
    # links to other nodes
    BinomialTree parent
    BinomialTree sibling
    BinomialTree child
```

For the BinomialHeap itself, we only need a reference to the first tree in the forest. This simple pseudocode is presented below:

```
class BinomialHeap
    BinomialTree head
```

As we progress toward a complete implementation, we will build up the operation that we need, working toward the union operation. Once union is implemented, adding or removing items from the heap can be implemented through the clever usage of union. For now, let's implement the combine and merge functions.

Combining Two Binomial Trees

The combine function is given below. This simple function combines two B_{k-1} trees to create a new B_k tree. This function will make the first tree a child of the second tree. We will assume that tree1.priority is always less than or equal to tree2.priority.

```
function combine(tree1, tree2)
    set tree1.parent to tree2
    set tree1.sibling to tree2.child
    set tree2.child to tree1
    set tree2.degree to tree2.degree + 1
```

The figure below shows how two B_2 trees would be combined to form a B_3 tree. This figure also identifies the parent, sibling, and child links for each node. Links that do not connect to any other node have the value null. This figure can help you understand the combine function. We need to maintain these links in a specific way to make the other operations function correctly. Notice that the children of the root are all linked together by sibling links, like a linked list's next reference.

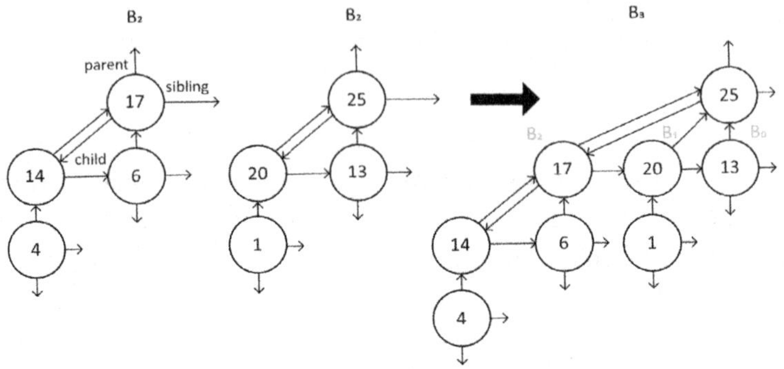

Figure 9.17

Another interesting feature of binomial trees is that each tree of degree k contains subtrees of all the degrees below it. These are the children linked from the first child of the new tree. For example, the B_3 tree contains a B_0, B_1, and B_2 subtree in descending order. You can observe these in the figure above. This fact will come in handy when we implement the extract function for binomial heaps.

Merging Heaps

Now that we can combine two trees to form a higher-degree tree, we should implement the mergeHeaps function. This will take both heaps and their forests of binomial trees and merge them. You can think of these "forests" as linked lists of binomial trees. Once these forests are merged, trees of equal degree will be adjacent in the list. This sets the stage for our "binary addition" algorithm that will calculate the union of both heaps. The code for mergeHeaps is below:

```
function mergeHeaps(heap1, heap2)
    set tree1 to heap1.head
    set tree2 to heap2.head
    set newHeap to an instance of BinomialHeap
    # if either heap is null, return the other
    if tree1 equals null
        return tree2
    if tree2 equals null
        return tree1

    # set the heap's head to the lowest degree tree
    if tree1.degree <= tree2.degree
        set newHeap.head to tree1
    else
        set newHeap.head to tree2
    set current to newHeap.head
    while (tree1 is not null) and (tree2 is not null)
        if tree1.degree <= tree2.degree
            # update sibling links
            set current.sibling to tree1
            set current to current.sibling
            # advance to tree1's next node
            set tree1 to tree1.sibling
        else
            # if this is not the first node, update sibling links
            set current.sibling to tree2
            set current to current.sibling
            # advance to tree2's next node
            set tree2 to tree2.sibling

    # tree1 or tree2 is null
    # end of one list was reached
    if tree1 equals null
        set current.sibling to tree2
    else
        set current.sibling to tree1

    return newHeap.head
```

This may look a little difficult to follow, but the concept is simple. Starting with links to the lists of binomial trees, we first check to see if any of these are null. If so, we just return the other one. Afterward, we check which of the nonnull trees has the lowest degree and set our newHeap's head to this tree. The while-loop then appends the tree with the next smallest degree to the growing list. The loop continues until one of the two lists of trees reaches its end. After that, the remaining trees are linked to the list by the current reference, and the head of our merged list is returned. Notice that the sibling references serve the same role as the next pointers of a linked list.

The Binomial Heap Union Operation

Now we will tackle the union function. This function performs the final step of combining two binomial heaps by traversing the merged list and combining any pairs of trees that have equal degrees. The algorithm is given below:

```
function union(heap1, heap2)
    set newHeap to an instance of BinomialHeap
    set newHeap.head to mergeHeaps(heap1, heap2)
    if newHeap.head equals null
        return newHeap

    # create references to traverse the merged list
    set previousTree to null
    set currentTree to newHeap.head
    set nextTree to newHeap.sibling

    while nextTree is not null
        # advance the traversal cases
        if (currentTree.degree is not nextTree.degree) or
        ((nextTree.sibling is not null) and
        (currentTree.degree equals nextTree.sibling.degree))
            set previousTree to currentTree
            set currentTree to nextTree
        else
           # combine the trees with equal degree by priority
           if currentTree.priority >= nextTree.priority
               set currentTree.sibling to nextTree.sibling
               # currentTree has higher priority
               combine(nextTree, currentTree)
           else
               # first node?  make sure to update the heap's head
               if previousTree equals null
                   set newHeap.head to nextTree
               else
                   set previousTree.sibling to nextTree
               # nextTree has higher priority
               combine(currentTree, nextTree)
        set nextTree to currentTree.sibling
    return newHeap
```

The union function begins by setting up a new, empty heap and then merging the two input heaps. Once the merged list of trees is generated, the algorithm traverses the list using three references (previousTree, currentTree, and nextTree). The code will advance the traversal forward if currentTree and nextTree have different degree values. Another case that moves the traversal forward is when nextTree.sibling has the same degree as currentTree and nextTree. This is because nextTree and nextTree.sibling will need to combine to occupy the k + 1 degree level. When a call to combine is needed, the algorithm checks which tree has the highest priority, and that tree's root becomes the root of the k + 1 tree.

Time Complexity of Union

The union operation is completed, but now we should analyze its complexity. One of the main reasons for choosing a binomial heap was its fast union operation. How fast is it though? We will be interested in the time complexity of union. The algorithm traverses both heaps for the merge and the sequence combinations. This means that the complexity is proportional to the number of trees. We need to determine how many binomial trees are needed to represent all n elements of the priority queue. Recall that there are many parallels between binomial heaps and binary numbers. If we have 3 items in our heap, we need a tree of degree 0 with 1 item and a tree of degree 1, with 2 items. To store 5 items, we would need a degree 0 tree with 1 item, and a degree 2 tree with 4 items. Here we see that the binary representation of n indicates which trees of any given degree are needed to store those elements. A number can be represented in binary using a maximum number of bits proportional to the log of that number. So there are log n trees in a binomial heap with n elements. This means that the time complexity of union is bounded by O(log n). By similar reasoning, the time cost for finding the element with the highest priority is O(log n), the number of binomial trees in the heap.

Inserting into a Binomial Heap

With union completed, we can see the benefit of this operation. Below is an implementation of insert using union. The union operation takes O(log n) time, and all other operations can be performed in O(1) time. This makes the time complexity for insert O(log n).

```
function insert(heap, data, priority)
    # create a new heap
    set newHeap to an instance of BinomialHeap

    # create a tree node
    set tree to an instance of BinomialTree
    set tree.data to data
    set tree.priority to priority
    set tree.degree to 0
    set tree.parent to null
    set tree.sibling to null
    set tree.child to null

    # add degree 0 tree to heap
    set newHeap.head to tree

    # union the heap and the new heap to insert
    return union(heap, newHeap)
```

Extracting the Max-Priority Element

The priority queue would not be complete without a function to extract the maximum element from the heap. The extract function takes a bit more work, but ultimately extract executes in O(log n) time. The maximum priority element must be the root of one of the heap's trees. This function will find the maximum priority element and remove its entire tree from the heap's tree list. Next, the children of the max-element are inserted into another heap. With the two valid heaps, we can now call union to create the binomial heap that results from removing the highest priority element. This element can be returned, and the binomial heap will have been updated to reflect its new state. We note that in this implementation, there is a side effect of extract. This function returns the maximum priority element and removes it from the input heap as a side effect that modifies the input. Another approach would be to implement an accessMax function to find the maximum priority element and return it without updating the heap. This means that extracting the element would require a call to accessMax to save the element, and then extract would be called.

We could also consider extracting part of the BinomialHeap class and avoid passing any heap as input.

```
function extract(heap)
    # first, find tree with max item
    set previousTree to null
    set currentTree to heap.head
    if currentTree equals null
        return null

    set maxTree to currentTree
    set maxPreviousTree to previousTree

    while currentTree is not null
        if currentTree.priority > maxTree.priority
            set maxTree to currentTree
            set maxPreviousTree to previousTree
        set previousTree to currentTree
        set currentTree to currentTree.sibling

    # connect previous tree to next tree in the heap
    if maxPreviousTree equals null
        # max is the first tree
        set heap.head to maxTree.sibling
    else
        set maxPreviousTree.sibling to maxTree.sibling

    # make a new heap for maxTree's children
    set newHeap to an instance of BinomialHeap
    set currentTree to maxTree.child

    # reverse and add children to the new heap
    while currentTree is not null
        # save the next node
        set nextTree to currentTree.sibling
        # add trees to the front of the list, reverse the order
        set currentTree.sibling to newHeap.head
        set currentTree.parent to null
        set newHeap.head to currentTree
        set currentTree to nextTree

    # now union modified heap and the heap of the max element's children
    set heap to union(heap, newHeap)
    return maxTree
```

Increase-Priority and Delete Operations

We will continue our theme of building new operations by combining old ones. Here we will present the delete operation. This operation will make use of extract and increasePriority, which we will develop next. Creating the increasePriority function will rely on the parent points that we have been maintaining. When the priority of an element is increased, it may need to work its way up the tree structure toward the root. The following code gives an implementation of increasePriority. We assume for simplicity that we already have a link to the element whose priority we want to increase. The element with the highest priority moves up the tree just like in the binary heap's siftUp operation.

```
function increasePriority(heap, tree, newPriority)
    if tree.priority > newPriority
        # do nothing if the new priority is less than the old.
        return

    set tree.priority to newPriority
    set currentTree to tree
    set parentTree to tree.parent
    while (parentTree is not null) and (currentTree.priority > parentTree.priority)
        # exchange the data and priority of the current and parent trees
        set temporaryData to parentTree.data
        set parentTree.data to currentTree.data
        set currentTree.data to temporaryData

        set temporaryPriority to parentTree.priority
        set parentTree.priority to currentTree.priority
        set currentTree.priority to temporaryPriority

        # move references up the tree toward the root
        set currentTree to parentTree
        set parentTree to parentTree.parent
```

Now we can implement delete very easily using the MAX special value. We increase the element's priority to the maximum possible value and then call extract. An implementation of delete is provided below. Again, we assume that a link to the element we wish to delete is provided.

```
function delete(heap, tree)
    increasePriority(heap, tree, MAX)
    extract(heap)
```

The time complexity of delete is derived from the complexity of increasePriority and extract. Each of these requires O(log n) time.

A Note on the Name

Before we close this chapter on priority queues, we should discuss why these are called binomial heaps. The name "binomial heap" comes from a property that for a binomial tree of degree k, the number of nodes at a given depth d is k *choose* d. This is written as $(k\ d) = \frac{k!}{d!(k-d)!}$.

Summary

In this chapter, we explored one of the most important abstract data types: the priority queue. This data structure provides a collection that supports efficient insert and remove operations with the added benefit of removing elements in order of priority. We looked at two important implementations. One implementation used an array and created a binary heap. We also saw that this structure can be used to implement an in-place O(n log n) sorting algorithm by simply extracting elements from the queue and placing them in the sorted zone of the array. Next, we explored the binomial heap's implementation. This data structure provides an O(log n) union operation. Though the implementation of binomial heaps seems complex, there is an interesting and beautiful simplicity to its structure. Our union algorithm also draws parallels to the addition of binary numbers, making it an interesting data structure to study. Any student of computer science should understand the concept of priority queues. They form the foundation of some interesting algorithms. For example, chapter 11 on graphs will present an algorithm for a minimum spanning tree that relies on an efficient priority queue. The minimum spanning tree is the backbone of countless interesting and useful algorithms. Hopefully, by now, you are beginning to see how data structures and algorithms build on each other through composition.

Exercises

1. Implement Heap Sort on arrays in the language of your choice. Revisit your work from chapter 3 on sorting. Using your testing framework, compare Heap Sort, Merge Sort, and Quick Sort. Which seems to perform better on average in terms of speed? Why would this be the case on your machine?
2. What advantages would Heap Sort have over Quick Sort? What advantages would Heap Sort have over Merge Sort?
3. Extend your heap implementation to use references to data records with a priority variable rather than just an integer as the priority.
4. Implement both binomial heaps using linked structures and binary heaps using an array implementation. Using a random number generator, create two equal-sized heaps, and try to merge them. Try the merge with binomial heaps, and get some statistics for the merge speed. Compare with the merge speed of binary heaps with integers. Repeat this process several times to explore how larger n affects the speed of the rival heap merge algorithms. The results may not be what you expect. How might caches play a role in the speed of these algorithms?

References

Cormen, Thomas H., Charles E. Leiserson, Ronald L. Rivest, and Clifford Stein. *Introduction to Algorithms*, 2nd ed. Cambridge, MA: The MIT Press, 2001.

Vuillemin, Jean. "A Data Structure for Manipulating Priority Queues." *Communications of the* ACM 21, no. 4 (1978): 309–315.

10. Dynamic Programming

Learning Objectives

After reading this chapter you will...

- understand the relationship between recursion and dynamic programming.
- understand the benefits of dynamic programming for optimization.
- understand the criteria for applying dynamic programming.
- be able to implement two classic dynamic programming algorithms.

Introduction

Dynamic programming is a technique for helping improve the runtime of certain optimization problems. It works by breaking a problem into several subproblems and using a record-keeping system to avoid redundant work. This approach is called "dynamic programming" for historical reasons. Richard Bellman developed the method in the 1940s and needed a catchy name to describe the mathematical work he was doing to optimize decision processes. The name stuck and, perhaps, leads to some confusion. This is because many terms in computer science have several meanings depending on the context, especially the terms "dynamic" and "programming." In any case, the technique of dynamic programming remains a powerful tool for optimization. Let's look deeper into this concept by exploring its link to problems that can be expressed recursively.

Recursion and Dynamic Programming

Recursive algorithms solve problems by breaking them into smaller subproblems and then combining them. Solving the subproblems is done by applying the same recursive algorithm to the smaller subproblems by breaking the subproblems into sub-subproblems. This continues until the base case is reached. Below is a recursive algorithm from chapter 2 for calculating the Fibonacci numbers:

```
function fibonacci(n)
  if n equals 0
    return 0
  else if n equals 1
    return 1
  else
    return fibonacci(n - 1) + fibonacci(n - 2)
```

To solve the problem for fibonacci(n), we need to solve it for fibonacci(n − 1) and fibonacci(n − 2). We see that there are subproblems with the same structure as the original problem.

The Fibonacci numbers algorithm is not an optimization problem, but it can give us some insight to help understand how dynamic programming can help us. Let's look at a specific instance of this problem. The recursive formula for Fibonacci numbers is given below:

$$F_0 = 0$$

$$F_1 = 1$$

$$F_n = F_{n-1} + F_{n-2}.$$

Now let's explore calculating the eighth Fibonacci number:

$$F_8 = F_{8-1} + F_{8-2}$$

$$= F_7 + F_6$$

$$= (F_{7-1} + F_{7-2}) + (F_{6-1} + F_{6-2})$$

$$= (F_6 + F_5) + (F_5 + F_4)$$

$$= ((F_{6-1} + F_{6-2}) + (F_{5-1} + F_{5-2})) + ((F_{5-1} + F_{5-2}) + (F_{4-1} + F_{4-2}))$$

$$= ((F_5 + F_4) + (F_4 + F_3)) + ((F_4 + F_3) + (F_3 + F_2))$$

...and so on.

There are two key thoughts we can learn from this expansion for calculating F_8. The first thought is that things are getting out of hand and fast! Every term expands into two terms. This leads to eight rounds of doubling. Our complexity looks like $O(2^n)$, which should be scary. Already at $n = 20$, 2^{20} is in the millions, and it only gets worse from there. The second thought that comes to mind in observing this explanation is that many of these terms are repeated. Let's look at the last line again.

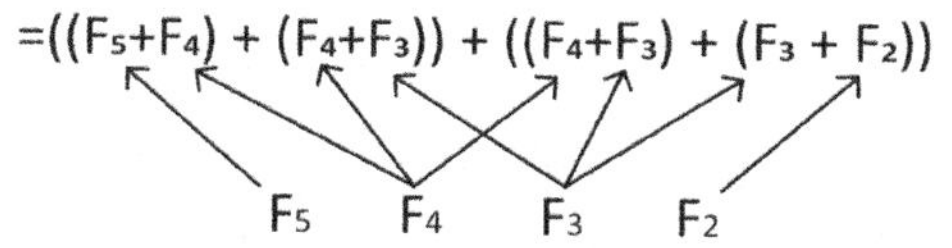

Figure 10.1

Already, we see that F_4 and F_3 are used three times each, and they would also be used in the expansion of F_5 and F_4. If we could calculate each of these just once and reuse the value, a lot of computation could be saved. This is the big idea of dynamic programming.

In dynamic programming, a record-keeping system is employed to avoid recalculating subproblems that have already been solved. This means that for dynamic programming to be helpful, subproblems must share sub-subproblems. In these cases, the subproblems are not independent of one another. There are some repeated identical structures shared by multiple subproblems. Not all recursive algorithms satisfy this property. For example, sorting one-half of an array with Merge Sort does not help you sort the other half. With Merge Sort, each part is independent of the other. In the case of Fibonacci, F_7 and F_6 both share a need to calculate F_1 through F_5. Storing these values for reuse will greatly improve our calculation time.

Requirements for Applying Dynamic Programming

There are two main requirements for applying dynamic programming. First, a problem must exhibit the property known as **optimal substructure**. This means that an optimal solution to the problem is constructed from optimal solutions to the subproblems. We will see an example of this soon. The second property is called **overlapping subproblems**. This means that subproblems are shared. We saw this in our Fibonacci example.

Optimal Matrix Chain Multiplication

A classic application of dynamic programming concerns the optimal multiplication order for matrices. Consider the sequence of matrices $\{M_1, M_2, M_3, M_4\}$. There are several ways to multiply these together. These ways correspond to the number of distinct ways to parenthesize the matrix multiplication order. For the mathematically curious, the Catalan numbers give the total number of possible ways. For example, one way to group these would be $(M_1 M_2) (M_3 M_4)$. Another way could be $M_1 ((M_2 M_3) M_4)$. Any grouping leads to the same final result, but the number of multiply operations of the overall calculation could differ greatly with different groupings. To understand this idea, let's review matrix multiplication.

Matrix Multiplication Review

Matrix multiplication is an operation that multiplies and adds the rows of one matrix with the columns of another matrix. Below is an example:

$$\underset{2 \times 3}{\overset{A}{\begin{bmatrix} a_{11} & a_{12} & a_{13} \\ a_{21} & a_{22} & a_{23} \end{bmatrix}}} \underset{3 \times 2}{\overset{B}{\begin{bmatrix} b_{11} & b_{12} \\ b_{21} & b_{22} \\ b_{31} & b_{32} \end{bmatrix}}} = \underset{2 \times 2}{\overset{C}{\begin{bmatrix} (a_{11}*b_{11} + a_{12}*b_{21} + a_{13}*b_{31}) & (a_{11}*b_{12} + a_{12}*b_{22} + a_{13}*b_{32}) \\ (a_{21}*b_{11} + a_{22}*b_{21} + a_{23}*b_{31}) & (a_{21}*b_{12} + a_{22}*b_{22} + a_{23}*b_{32}) \end{bmatrix}}}$$

Figure 10.2

Here we have the matrix A and the matrix B. A is a 2-by-3 matrix (2 rows and 3 columns), and B is a 3-by-2 matrix (3 rows and 2 columns). The multiplication of AB is compatible, which means the number of columns of A is equal to the number of rows in the second matrix, B. When two compatible matrices are multiplied, their result has a structure where the number of rows equals the number of rows in the first matrix and the number of columns equals the number of columns in the second matrix. The process is the same for compatible matrices of any size.

Implementing Matrix Multiplication

Now let's consider an algorithm for matrix multiplication. To simplify things, let's assume we have a Matrix class or data structure that has a two-dimensional (2D) array. Another way to think of a 2D array is as an array of arrays. We could also think of this as a table with rows and columns. The structure below gives a general example of a Matrix class. Within this class, we also have two convenience functions to access and set the values of the matrix based on the row and column of the 2D array.

```
class Matrix
    Integer rows
    Integer columns
    2D array data

    function at(r, c)
        return data[r][c]

    function set(r, c, value)
        set data[r][c] to value
```

With this structure for a Matrix class, we can implement a matrix multiplication procedure. Below we show the process of performing matrix multiplication on two compatible matrices:

```
1  function matrixMultiply(matrixA, matrixB)
2      create Matrix matrixC
3      set matrixC.rows to matrixA.rows
4      set matrixC.columns to matrixB.columns
5      for i from 0 to matrixA.rows − 1
6          for j from 0 to matrixB.columns − 1
7              # calculate entry of C[i][j]
8              set value to 0
9              for k from 0 to matrixA.columns − 1
10                 set valueA to matrixA.at(i, k)
11                 set valueB to matrixB.at(k, j)
12                 set value to value + valueA*valueB
13             # set the output matrixC's value
14             matrixC.set(i, j, value)
15     return matrixC
```

This function implements the matrix multiplication procedure described above. On line 12, the new value of the (i, j) entry in the result matrix is calculated. We see that this involves a multiply operation and an addition operation. On a typical processor, the multiply operation is slower than addition. As we think about the complexity of matrix multiplication, we will mainly consider the number of multiplications. This is because as the matrices get large, the cost associated with multiplication will dominate the cost of addition. For this reason, we only consider the number of multiplications.

So how many multiplications are needed for matrix

multiplication? The pattern above has a triple-nested loop. This gives us a clue to the number of times the inner code will run. As a result, we can expect the number of multiplications to be equal to the number of times the inner code will run. Let's assume that matrix A has r_a rows and c_a columns, and that matrix B, in a similar way, has r_b rows and c_b columns. For A and B to be compatible matrices, the value of c_a would have to be equal to r_b. We know that the inner loop with index k runs a total of c_a times. This entire loop is executed once for every c_b of B's columns ($c_b * c_a$). Finally, these two inner loops for j and k would all run for every row in A, leading to multiplications proportional to $r_a * c_a * c_b$. This illustrates that as the size of the matrices gets larger, the number of multiplications grows quickly.

Why Order Matters

Now that we have seen how to multiply matrices together and understand the computational cost, let's consider just why choosing to multiply in a specific order is important. Suppose that we need to multiply three matrices—A, B, and C—shown in the image below:

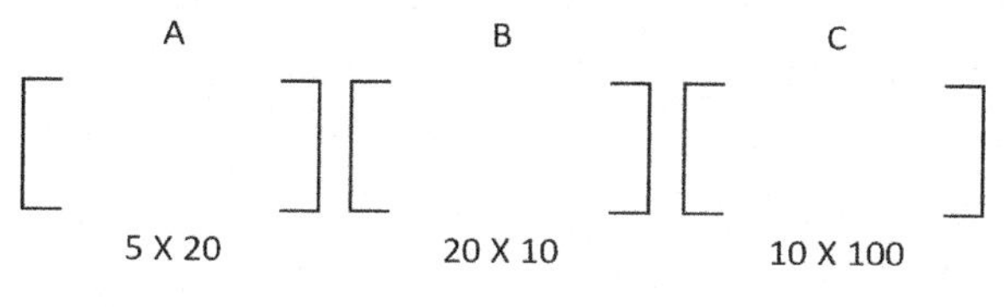

Figure 10.3

Multiplying them together could proceed with the

grouping (AB)C, where A and B are multiplied together first, and then that result is multiplied by C. Alternatively, we could group them as A(BC) and first multiply B by C, followed by A multiplied by the result. Which would be better, or would it even matter?

Let's figure this out by first considering the A(BC) grouping. The figure below illustrates this example. With this grouping, calculating the BC multiplication yields 20,000 multiply operations. Multiplying A by this result gives another 10,000 for a total of 30,000 multiply operations.

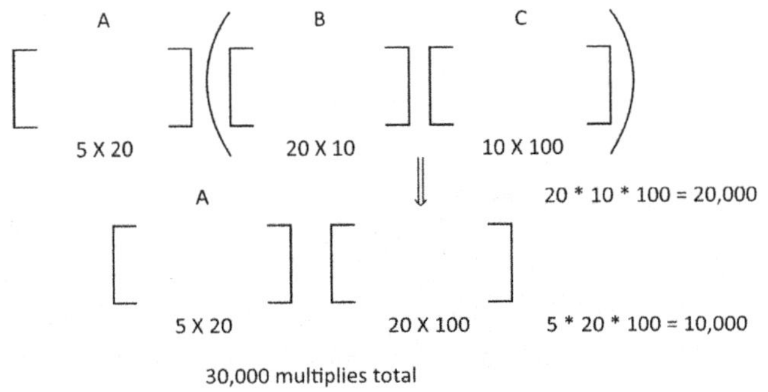

Figure 10.4

Next, let's consider the (AB)C grouping. The following figure shows a rough diagram of this calculation. The AB matrix multiplication gives a cost of 1,000 multiply operations. Then this result multiplied by C gives another 5,000. We now have a total of 6,000 multiply operations for the (AB)C grouping over the other. This represents a fivefold decrease in cost!

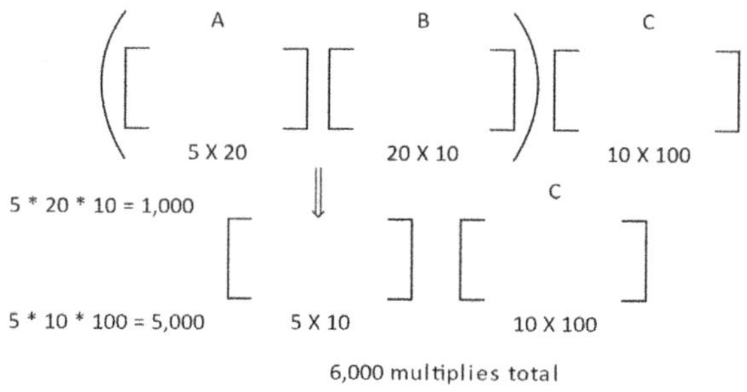

Figure 10.5

This example illustrates that the order of multiplication definitely matters in terms of computational cost. Additionally, as the matrices get larger, there could be significant cost savings when we find an optimal grouping for the multiplication sequence.

A Recursive Algorithm for Optimal Matrix-Chain Multiplication

We are interested in an algorithm for finding the optimal ordering of matrix multiplication. This corresponds to finding a grouping with a minimal cost. Suppose we have a chain of 5 matrices, M_0 to M_4. We could write their dimensions as a list of 6 values. The 6 values come from the fact that each sequential pair of matrices must be compatible for multiplication to be possible. The figure below shows this chain and gives the dimensions as a list.

M_0 M_1 M_2 M_3 M_4

$d_0 \times d_1$ $d_1 \times d_2$ $d_2 \times d_3$ $d_3 \times d_4$ $d_4 \times d_5$

dimensions = $[d_0, d_1, d_2, d_3, d_4, d_5]$

Figure 10.6

An algorithm that minimizes the cost must find an optimal split for the final two matrices. Let's call these final two matrices A and B. For the result to be optimal, then A and B must both have resulted from an optimal subgrouping. The possible splits would be

$(M_0)(M_1 M_2 M_3 M_4)$ = AB with a split after position 0

$(M_0 M_1)(M_2 M_3 M_4)$ = AB with a split after position 1

$(M_0 M_1 M_2)(M_3 M_4)$ = AB with a split after position 2

$(M_0 M_1 M_2 M_3)(M_4)$ = AB with a split after position 3.

We need to evaluate these options by assessing the cost of creating the A and B matrices (optimal subproblems) as well as the cost of the final multiply, with matrix A being multiplied by B. A recursive algorithm would find the minimal cost by checking the minimum cost among all splits. In the process of finding the cost of all these four options for splits, we would need to calculate the optimal splits for other sequences to find their optimal groupings. This demonstrates the feature of **optimal substructure**, the idea that an optimal solution could be built from optimal subproblems.

For the first grouping, we have A = M_0 and B = (M_1 M_2 M_3 M_4). To calculate the cost of this split, it is assumed that A and B have been constructed optimally. This means that a recursive algorithm considering this split must then make a recursive call to find the minimal grouping for (M_1 M_2 M_3 M_4) for the B matrix. This in turn would trigger another search for the optimal split among (M_1) (M_2 M_3 M_4) (M_1 M_2) (M_3 M_4) and (M_1 M_2 M_3) (M_4). We can also see that this would trigger further calls to optimize each sequence of 3 matrices and so on. You may be able to imagine that this recursive process has a high branch factor leading to an exponential runtime complexity in the number of matrices. With n matrices, the runtime complexity would be even worse than $O(2^n)$, **exponential** time. It would follow an algorithm for calculating the Catalan numbers at $O(3^n)$.

A general outline of the recursive algorithm would be as follows. We will consider an algorithm to calculate the minimal cost of multiplying a sequence of matrices starting at some matrix identified by the start index and including the ending matrix using an end index. The base case of the algorithm is when start and end are equal. The cost of multiplication of only one matrix is 0 as there is no operation to perform. The recursive case calculates the cost of splitting the sequence at some split position. There will be n – 1 split positions to test when considering n matrices where n = end – start + 1, and the recursive algorithm will need to find the minimum of the options for the best split position.

To complete the recursive algorithm, we will introduce a function to calculate the cost of the final multiplication. This could be a simple multiplication of the correct dimensions, but we will introduce and explain this function to make the meaning clear and to simplify some of the code (which would otherwise include a lot of awkward indexing). The figure below illustrates what is meant by the final multiplication:

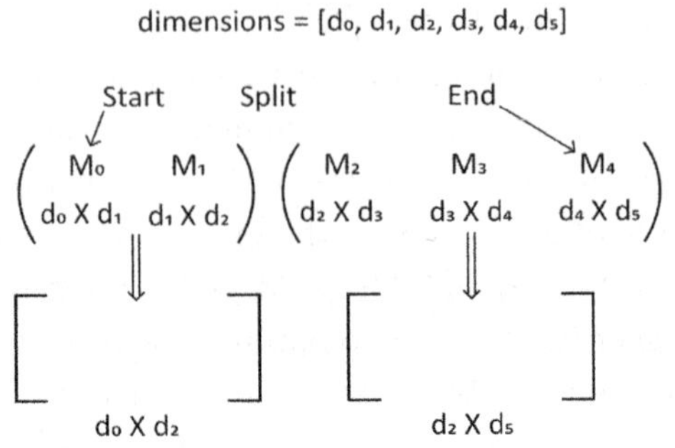

Figure 10.7

Suppose we are calculating the number of multiplications for a split at index 1 (or just after M_1). The algorithm would have given the optimal cost for constructing the left matrix and the right matrix, but we would still need to calculate the cost of multiplying those together. The left matrix would have dimensions of d_0 by d_2 and the right matrix would have dimensions of d_2 by d_5. Using the dimensions list and indexes for start, split, and end, we can calculate this cost. The function below performs this operation in a way that might make the meaning a little clearer. Notice that for matrix i, the dimensions of that matrix are d_i by d_{i+1}.

```
function costOfLastMultiply(dimensions, start, split, end)
    set dimension1 to dimensions[start]
    set dimension2 to dimensions[split + 1]
    set dimension3 to dimensions[end + 1]
    return dimension1 * dimension2 * dimension3
```

With this helper function, we can now write the recursive algorithm.

```
function recursiveChainOpt(dimensions, start, end)
    if start equals end
        return 0
    else
        set currentMin to MAX
        for splitIndex from start to end - 1
            set leftCost to recursiveChainOpt(dimensions, start, splitIndex)
            set rightCost to recursiveChainOpt(dimensions, splitIndex + 1, end)
            set lastCost to costOfLastMultiply(dimensions, start, splitIndex, end)
            set currentCost to leftCost + rightCost + lastCost
            if currentCost < currentMin
                set currentMin to currentCost
        return currentMin
```

This algorithm only calculates an optimal cost, but it could be modified to record the split indexes of the optimal splits so that another process could use that information. The optimal cost of multiplying all matrices in the optimal grouping could be calculated with a call to recursiveChainOpt(dimensions, 0, 4). This algorithm, while correct, suffers from exponential time complexity. This is the type of situation where dynamic programming can help.

A Dynamic Programming Solution

Let's think back to our precious example for a moment. Think specifically about the first two groupings we wanted to consider. These are (M_0) (M_1 M_2 M_3 M_4) and (M_0 M_1) (M_2 M_3 M_4). For the first grouping, we need to optimize the grouping of (M_1 M_2 M_3 M_4) as a subproblem. This would involve also considering the optimal grouping of (M_2 M_3 M_4). Optimally grouping (M_2 M_3 M_4) is a problem that must be solved in the process of calculating the cost of (M_0 M_1) (M_2 M_3 M_4), which is the second subproblem in the original grouping. From this, we see that there are **overlapping subproblems**. Considering this problem meets the criteria of optimal substructure and overlapping subproblems, we can be confident that dynamic programming will give us an advantage.

The logic behind the dynamic programming approach is to calculate the optimal groupings for subproblems first, working our way through larger and larger subsequences and saving their optimal cost. Eventually, the algorithm minimizes the cost of the full sequence of matrix multiplies. In this calculation, the algorithm queries the optimal costs of the smaller sequences from a table. This algorithm uses two tables. The first table, modeled using a 2D array, stores the calculated optimal cost of multiplying matrices i through j. This table will be called costs. The second table holds the choice of split index associated with the optimal cost. This table will be called splits. While the algorithm calculates costs, the splits are the important data that can be used to perform the actual multiplication in the right order.

The algorithm is given below. It begins by assigning the optimal values for a single matrix. A single matrix has no multiplies, so when calculating a matrix chain multiplication with a sequence of 1, the cost is 0. Next, the algorithm sets a sequence length starting at 2. From here, start and end indexes are set and updated such that the optimal cost of all length 2 sequences in the chain are calculated and stored in the costs table. Next, the sequence length is increased to 3, and all optimal sequences of length 3 are calculated by trying the different options for the splitIndex. The splitIndex is updated in the splits table each time an improvement in cost is found. We should note that we again make use of the MAX value, which acts like infinity as we minimize the cost for a split. The process continues for larger and larger sequence lengths until it finds the cost of the longest sequence, the one including all the matrices.

```
function optimalMultiplyGrouping(dimensions)
    set n to length(dimensions) - 1
    create 2D-array costs of size n-by-n
    create 2D-array splits of size n-by-n
    for index from 0 to n - 1
        # a single matrix requires no cost
        set costs[index][index] to 0
    # consider optimal groupings for different subsequences
    for subsequenceLength from 2 to n
        set lastStart to n - subsequenceLength
        for start from 0 to lastStart
            set end to start + subsequenceLength - 1
            set cost[start][end] to MAX
            for splitIndex from start to end - 1
                set multiplyCost to costOfLastMultiply(dimensions, start, split, end)
                set leftCost to costs[start][splitIndex]
                set rightCost to costs[splitIndex + 1][end]
                set currentCost to leftCost + rightCost + multiplyCost
                if currentCost < costs[start][end]
                    set costs[start][end] to currentCost
                    set splits[start][end] to splitIndex
    # return the table that gives the optimal split positions
    return splits
```

Complexity of the Dynamic Programming Algorithm

Now we have seen two algorithms for solving the optimal matrix chain multiplication problem. The recursive formulation proved to be exponential time ($O(2^n)$) with each recursive call potentially branching n - 1 times. The dynamic programming algorithm should improve upon this cost; otherwise, it would not be very useful. One way to reason about the complexity is to think about how the tables get filled in. Ultimately, we are filling in about one-half of a 2D array or table. This amounts to filling in the upper triangular portion of a matrix in mathematical terms. Our table is n by n, and we are filling in n(n+1)/2 values (a little over half of the n-by-n matrix). So you may think the time complexity should be $O(n^2)$. This is not the full story though. For every start-end pair, we must try all the split indexes. This could be as bad as n - 1. So all these pairs need to evaluate up to n - 1 options for a split. We can reason that this requirement would lead to some multiple of n^2*n or n^3 operations. This provides a good explanation of the time complexity, which is $O(n^3)$. This may seem expensive, but $O(n^3)$ is profoundly better than $O(3^n)$. Moreover, consider the difference between the number matrices and the number multiplications needed for the chain multiplication. For our small example of 3 matrices (our n in this case), we saw the number of multiply operations drop by 24,000 when using the optimal grouping, and our n was only 3. This could result in a significant improvement in the overall computation time, making the optimization well worth the cost.

Longest Common Subsequence

Another classic application of dynamic programming involves detecting a shared substructure between two strings. For example, the two strings "pride" and "ripe" share the substring "rie." For these two strings, "rie" is the longest common subsequence or LCS. There are other subsequences, such as "pe," but "rie" is the longest or optimal subsequence. These subsequence strings do not need to be connected. They can have nonmatched characters in between. It might seem like a fair question to ask, "Why is this useful?" Finding an LCS may seem like a simple game or a discrete mathematics problem without much significance, but it has been applied in the area of computational biology to perform alignments of genetic code and protein sequences. A slight modification of the LCS algorithm we will learn here was developed by Needleman and Wunsch in 1970. That algorithm inspired many similar algorithms for the dynamic alignment of biological sequences, and they are still empowering scientific discoveries today in genetics and biomedical research. Exciting breakthroughs can happen when an old algorithm is creatively applied in new areas.

Defining the LCS and Motivating Dynamic Programming

A common subsequence is any shared subsequence of two strings. A subsequence of a string would be any ordered subset of the original sequence. An LCS just requires that this be the longest such subsequence belonging to both strings. We say "an" LCS and not "the" LCS because there could be multiple common subsequences with the same optimal length.

Let's add some terms to better understand the problem. Suppose we have two strings A and B with lengths m and n, respectively. We can think of A as a sequence of characters $A = \{a_0, a_1, ..., a_{m-1}\}$ and B as a sequence of the form $B = \{b_0, b_1, ..., b_{n-1}\}$. Suppose we already know that C is an LCS of A and B. Let's let k be the length of C. We will let A_i or B_i mean the subsequence up to i, or $A_i = \{a_0, a_1, ..., a_i\}$. If we think of the last element in C, C_{k-1} must be in A and B. For this to be the case, one of the following must be true:

1. $c_{k-1} = a_{m-1}$ and $c_{k-1} = b_{n-1}$. This means that $a_{m-1} = b_{n-1}$ and C_{k-2} is an LCS of A_{m-2} and B_{n-2}.
2. a_{m-1} is not equal to b_{n-1}, and c_{k-1} is not equal to a_{m-1}. This must mean that C is an LCS of A_{m-2} and B.
3. a_{m-1} is not equal to b_{n-1}, and c_{k-1} is not equal to b_{n-1}. This must mean that C is an LCS of A and B_{n-2}.

In other words, if the last element of C is also the last element of A and B, then it means that the subsequence C_{k-2} is an LCS of A_{m-2} and B_{n-2}. This is hinting at the idea of **optimal substructure**, where the full LCS could be built from the C_{k-2} subproblem. The other two cases also imply subproblems where an LCS, C, is constructed from either the case of A_{m-2} (A minus its last element) and B or the case of A and B_{n-2} (B minus its last element).

Now let's consider **overlapping subproblems**. We saw that our optimal solution for an LCS of A and B could be built from an

LCS of A_{m-2} and B_{n-2} when the last elements of A and B are the same. Finding an LCS of A_{m-2} and B_{n-2} would also be necessary for our other two cases. This means that in evaluating which of the three cases leads to the optimal LCS length, we would need to evaluate the LCS of A_{m-2} and B_{n-2} subproblems and potentially many other shared problems with shorter subsequences. Now we have some motivation for applying dynamic programming with these two properties satisfied.

A Recursive Algorithm for Longest Common Subsequence

Before looking at the dynamic programming algorithm, let's consider the recursive algorithm. Given two sequences as strings, we wish to optimize for the length of the longest common subsequence. The algorithm below provides a recursive solution to the calculation of the optimal length of the longest common subsequence. Like with our matrix chain example, we could add another list to hold each element of the LCS, but we leave that as an exercise for the reader.

```
function recursiveLCS(stringA, stringB, indexA, indexB)
    if indexA equals -1 or indexB equals -1
        return 0
    else
        if stringA[indexA] equals stringB[indexB]
            return 1 + recursiveLCS(stringA, stringB, indexA - 1, indexB - 1)
        else
            set shortenALength to recursiveLCS(stringA, stringB, indexA - 1, indexB)
            set shortenBLength to recursiveLCS(stringA, stringB, indexA, indexB - 1)
            if shortenALength > shortenBLength
                return shortenALength
            else
                return shortenBLength
```

This algorithm will report the optimal length using a function call with the last valid indexes (length – 1) of each string provided as the initial index arguments. For example, letting A be "pride" and B be "ripe," our call would be recursiveLCS(A, B, 4, 3), with 4 and 3 being the last valid indexes in A and B. Trying to visualize the call sequence for this recursive algorithm, we could image a tree structure that splits into two branches each time the else block is executed on line 7. In the worst case, where there are no shared elements and the length of an LCS is 0, this means a new branch generates 2 more branches for every n elements (assuming n is larger than m). This leads to a time complexity of $O(2^n)$.

A Dynamic Programming Solution

In a similar way to the matrix chain algorithm, our dynamic programming solution for LCS makes use of a table to record the length of the LCS for a specific pair of string indexes. Additionally, we will use another "code" table to record from which optimal subproblem the current optimal solution was constructed.

The following dynamic programming solution tries to find LCS lengths for all subsequences of the input strings A and B. First, a table, or 2D array, is constructed with dimensions (n+1) by (m+1). This adds an extra row and column to accommodate the LCS of a sequence and an empty sequence or nothing. A string and the empty string can have no elements in common, so the algorithm initializes the first row and column to zeros. Next, the algorithm proceeds by attempting to find the LCS length of all subsequences of string A and the first element of string B. For any index pair (i, j), the algorithm calculates the LCS length for the two subsequence strings A_j and B_i.

The core of the algorithm checks the three cases discussed above. These are the case of a match among elements of A and B and two other cases where the problem could be reformulated as either shortening the A string by one or shortening the B string by one. As the algorithm decides which of these options is optimal, we record a value into our "code" table that tells us which of these options was chosen. We will use the code "D," "U," and "L" for "Diagonal," "from the Upper entry," and "from the Left entry." These codes will allow us to easily traverse the table by moving "diagonal," "up," or "left," always taking an optimal path to output an LCS string. Let's explore the algorithm's code and then try to understand how it works by thinking about some intermediate states of execution.

```
function longestCommonSubsequence(stringA, stringB)
    set m to length(stringA)
    set n to length(stringB)
    create 2D-array lengths of size (n+1)-by-(m+1)
    create 2D-array codes of size (n+1)-by-(m+1)
    # initialize first row and column in the table to 0
    for index from 0 to n
        set lengths[index][0] to 0
    for index from 0 to m
        set lengths[0][index] to 0

    for indexB from 1 to n
        for indexA from 1 to m
            # the string position is offset by 1 due to increased table size.
            if stringA[indexA-1] equals stringB[indexB-1]
                set previousLCSLength to lengths[indexB-1][indexA-1]
                set lengths[indexB][indexA] to 1 + previousLCSLength
                set codes[indexB][indexA] to "D"
            else
                set shortenA to lengths[indexB][indexA-1]
                set shortenB to lengths[indexB-1][indexA]
                if shortenA > shortenB
                    set lengths[indexB][indexA] to shortenA
                    set codes[indexB][indexA] to "L"
                else
                    set lengths[indexB][indexA] to shortenB
                    set codes[indexB][indexA] to "U"
    # return the table of optimal choice codes
    return codes
```

To better understand the algorithm, we will explore our previous example of determining the LCS of "pride" and "ripe." Let us imagine that the algorithm has been running for a bit and we are now examining the point where indexB is 2 and indexA is 4. The figure below gives a diagram of the current states of the lengths and codes tables at this point in the execution:

		0	p 1	r 2	i 3	d 4	e 5
	0	0	0	0	0	0	0
r	1	0	0	1	1	1	1
i	2	0	0	1	2	?	
p	3	0					
e	4	0					

lengths table

		0	p 1	r 2	i 3	d 4	e 5
	0	0	0	0	0	0	0
r	1	0	U	D	L	L	L
i	2	0	U	U	D	?	
p	3	0					
e	4	0					

codes table

Figure 10.8

These tables can give us some intuition on how the algorithm works. Looking at the lengths table in row 2 and column 2, we see there is a 1. This represents the LCS of the strings "pr" and "ri." They share a single element "r." Moving over to the cell found in row 2 and column 3, we see the number 2. This represents the length of the LCS of the strings "pri" and "ri." Now the algorithm is considering the cell in the row 2 column 4 position. The strings in these positions do not match, so this is not built from the LCS along the diagonal. The largest LCS value from the previous subproblems is 2. This means that the LCS of "prid" and "ri" is the same as the LCS of the shortened A string "pri" and "ri." Since this is the case, we would mark a 2 at this position in the lengths table and make an "L" in this position in the codes table. These algorithms take time to understand fully. Don't get discouraged if it doesn't click right away. Try to implement it in your favorite programming language, and work on some examples by hand. Eventually, it will become clear.

Extracting the LCS String

Before we move on to the complexity analysis, let's discuss how to read the LCS from the codes table. Depending on the design of your algorithm, you may be able to extract the LCS just from the strings and the lengths table, and the codes table could be omitted from the algorithm completely. We would like to keep things simple though, so we will just use the codes table. The algorithm below shows one method for printing the LCS string (in reverse order):

```
function printLCS(codes, stringA, rows, columns)
    set rowIndex to rows - 1
    set columnIndex to columns - 1
    while rowIndex > 0 and columnIndex > 0
        if codes[rowIndex][columnIndex] equals "D"
            # print the string element at columnIndex - 1
            print stringA[columnIndex-1]
            set rowIndex to rowIndex - 1
            set columnIndex to columnIndex - 1
        else
            if codes[rowIndex][columnIndex] equals "U"
                set rowIndex to rowIndex - 1
            else
                set columnIndex to columnIndex - 1
```

Complexity of the Dynamic Programming Algorithm

Now that we have seen the algorithm and an example, let's consider the time complexity of the algorithm. The nested loops for the A and B indexes should be a clue. In the worst case, all increasing subsequences of each input string need to be compared. The algorithm fills every cell of the n-by-m table (ignoring the first row and column of zeros, which are initialized with a minor time cost). This gives us n * m cells, so the complexity of the algorithm would be considered O(mn). It might be reasonable to assume that m and n are roughly equal in size. This would lead to a time complexity of $O(n^2)$. This represents a huge cost savings over the $O(2^n)$ time cost of the recursive algorithm.

The space complexity is straightforward to calculate. We need two tables, each of size n+1 by m+1. So the space complexity would also be O(m*n), or, assuming m is roughly equal to n, $O(n^2)$.

A Note on Memoization

A related topic often appears in discussions of dynamic programming. The main advantage of dynamic programming comes from storing the result of costly calculations that may need to be queried later. The technique known as memoization does this in an elegant way. One approach to memoization allows for a function to keep a cache of tried arguments. Each time the function is called with a specific set of arguments, the cache can be queried to see if that result is known. If the specific combination of arguments has been used before, the result is simply returned from the cache. If the arguments have not been seen before, the calculation proceeds as normal. Once the result is calculated, the function updates the cache before returning the value. This will save work for the next time the function is called with the same arguments. The cache could be implemented as a table or hash table.

The major advantage of memoization is that it enables the use of recursive style algorithms. We saw in chapter 2 that recursion represents a very simple and clear description of many algorithms. Looking back at the code in this chapter, much of it is neither simple nor clear. If we could have the best of both worlds, it would be a major advantage. Correctly implementing memoization means being very careful about variable scope and correctly updating the cache when necessary to make sure the optimal value is returned. You must also be reasonably sure that querying your cache will be efficient.

Summary

Dynamic programming provides some very important benefits when used correctly. Any student of computer science should be familiar with dynamic programming at least on some level. The most important point is that it represents an amazing reduction in complexity from exponential $O(2^n)$ to polynomial time complexity $O(n^k)$ for some constant k. Few if any other techniques can boast of such a claim. Dynamic programming has provided amazing gains in performance for algorithms in operations research, computational biology, and cellular communications networks.

These dynamic programming algorithms also highlight the complexity associated with implementing imperative solutions to a recursive problem. Often writing the recursive form of an algorithm is quite simple. Trying to code the same algorithm in an imperative or procedural way leads to a lot of complexity in terms of the implementation. Keeping track of all those indexes can be a big challenge for our human minds.

Finally, dynamic programming illustrates an example of the **speed-memory trade-off**. With the recursive algorithms, we only need to reserve a little stack space to store some current index values. These typically take up only O(log n) memory on the stack. With dynamic programming (including memoization styles), we need to store the old results for use later. This takes up more and more memory as we accumulate a lot of partial results. Ultimately though, having these answers stored and easily accessible saves a lot of computation time.

Exercises

1. Think of some other recursive algorithms that you have learned. Do any of them exhibit the features of optimal substructure and overlapping subproblems? Which do, and which do not?

2. Try to implement the dynamic programming algorithm for optimal matrix chain multiplication. Next, implement a simple procedure that calculates the cost of a naïve matrix multiplication order that is just a typical left-to-right multiplication grouping. Randomly generate lists of dimensions, and calculate the costs of optimal vs. naïve. What patterns do you observe? Are there features of matrix chains that imply optimizations?

3. Try to implement a recursive function to print the optimal parenthesization of the matrix multiplication chain given the splits table. **Hint:** Accept a start and end value, and for each split index s (splits[start][end]), recursively call the function for (start, s) and (s + 1, end).

4. Implement the recursive LCS algorithm in your language of choice, and extend it to report the actual LCS as a string. **Hint:** You may need to use a data structure to keep track of the current LCS elements.

5. Extend the LCS algorithm to implement an alignment algorithm for genetic code (strings containing only {"a," "c," "g," "t"} elements). This could be done by adding a scoring system. When the algorithm assesses a diagonal, check if it is a match (exact element) or a mismatch (elements are not the same). Calculate an optimal score using the following rules: Matches get +2,

mismatches get −1, moving left or up counts as a "gap" and gets −2. Calculate the optimal alignment score using this method for the strings "acctg" and "gacta."

References

Bellman, Richard. *Eye of the Hurricane*. World Scientific, 1984.

Cormen, Thomas H., Charles E. Leiserson, Ronald L. Rivest, and Clifford Stein. *Introduction to Algorithms*, 2nd ed. Cambridge, MA: The MIT Press, 2001.

Needleman, Saul B., and Christian D. Wunsch. "A General Method Applicable to the Search for Similarities in the Amino Acid Sequence of Two Proteins." *Journal of Molecular Biology* 48, no. 3 (1970): 443–453.

11. Graphs

Learning Objectives

After reading this chapter you will...

- understand graphs as a mathematical structure.
- be able to traverse graphs using well-known algorithms.
- learn the scope of problems that can be addressed using graphs.

Introduction

Graphs are perhaps the most versatile data structure addressed in this book. As discussed in chapter 8, graphs at their simplest are just nodes and edges, with nodes representing things and the edges representing the relationships between those things. However, if we are creative enough, we can apply this concept to the following:

- roads between cities
- computer networks
- business flow charts
- finite state machines
- social networks
- family trees

- circuit design

If we can represent a given problem using a graph, then we have access to many well-known algorithms to help us solve that problem. This chapter will introduce us to graphs as well as some of these algorithms.

Brief Introduction to Graphs

A formal introduction to graphs may be found in most discrete math textbooks. The purpose of this section is not to provide such an introduction. Rather, we will focus more on the data structures used to represent graphs and the algorithms associated with them. Regardless, some basic vocabulary will prove useful.

Nodes are the primary objects under consideration in graphs. They are associated with other nodes by means of edges. Each edge is incident to two nodes. We define adjacent nodes of a given node as alternate nodes of incident edges. Adjacent nodes are sometimes referred to as neighbors. The degree of a node is the number of its incident edges or adjacent nodes (not including the node itself). Our depiction of graphs will look much like our depiction of trees. This should come as no surprise because, if you recall, trees are a special case of graphs.

If you cross-reference this chapter against more mathematically focused textbooks (which is strongly encouraged), you will find that some of these definitions vary. In particular, mathematical textbooks typically refer to nodes as vertices. They are also more precise in mathematical notation. For example, this chapter will use the symbol N to denote the number of nodes in the graph. In a mathematics context, the set of nodes is represented as set V and the number of nodes as |V| (or the cardinality of V).

Paths and cycles will play a large role in our graph algorithms. A path is some sequence of edges that allows you to travel from one node to another. A cycle is a path that starts and ends at the same node. Sometimes cycles are useful, but they often represent challenges in graph algorithms. Failing to detect a cycle in graph algorithms often results in implementations falling into endless loops.

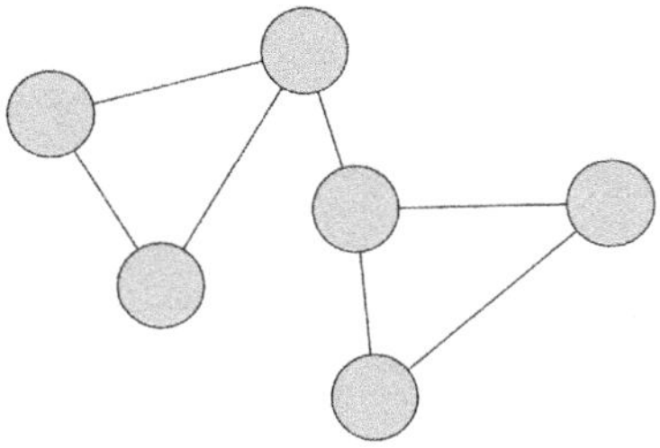

Figure 11.1

We will often need to consider whether a graph is directed or undirected. A directed graph is one in which each edge goes in a single direction. A network of flights across the United States is probably directed because every flight from city A to city B does not necessarily have a flight back from city B to city A. An undirected graph implies that we could traverse each edge in either direction. A network of roads between towns is likely undirected because most roads permit travel in both directions. If you reframe the network of roads, you could describe each lane as an edge, resulting in a directed graph. In our visual depictions of graphs, you will know whether a graph is directed or undirected by the use of arrows. Of the graphs below, the left graph represents an undirected graph, and the right represents a directed graph:

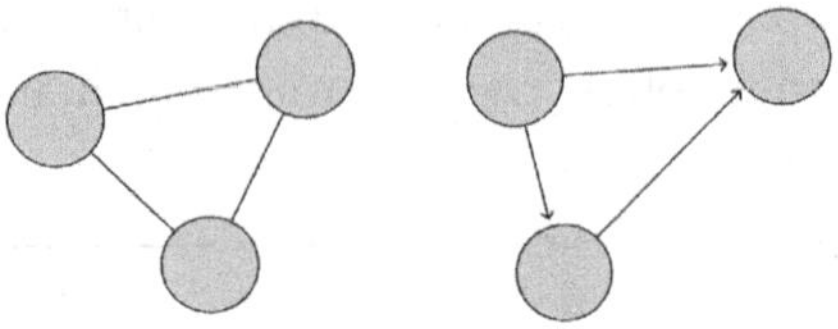

Figure 11.2

Another distinction we make will be between weighted and unweighted graphs. Weighted graphs have a numerical value associated with each edge, which represents the weight of that edge. For example, roads between cities have distances, and computer networks have measures of latency. Some graphs have no meaningful numerical value for edges and are considered unweighted. For example, social networks may have no meaningful weight for the relationship between two people. Much of this chapter will focus on graphs that are both weighted and directed.

Representations of Graphs

In a discrete math class, these graphs would be represented with basic set notation. We, however, have the additional burden of needing to represent this in a machine-readable format. Two main strategies exist for representing graphs in data structures, but there are numerous variations on these. We may choose to modify or augment these structures depending on the specific problem, language, or computing environment. For simplicity, we will only address the two main strategies.

Adjacency Matrices

An adjacency matrix is typically conceptualized as a table where the count of both rows and columns is equal to the number of nodes in the graph. Each row is assigned a node identifier, and each column is also assigned a node identifier. To determine whether an edge exists from node A to node B, we find the row for A and cross-reference B. The information in that cell of the table then provides information regarding the nature of the edge from A to B. Consider the following example:

	A	B	C
A	Exists	Exists	Exists
B			Exists
C			

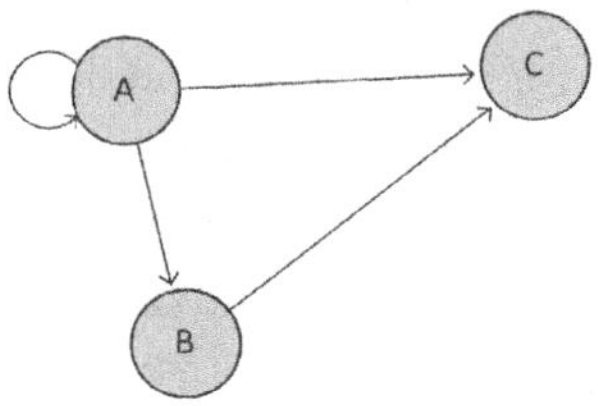

Figure 11.3

Notice that all we have asserted so far is that the intersection of the row and column supplies information about the nature of the edge. The data we store at each intersection depends on the type of graph we are modeling. Some considerations for adjacency matrices include the following:

- Weighted or Unweighted: If the graph is weighted, intersections will store the weight of the edge as some numeric type. In unweighted graphs, the intersection simply stores whether the edge exists.
- Directed or Undirected: If the graph is undirected, each nondiagonal intersection stores redundant information with exactly one other cell. For example, if an undirected graph has an edge between A and B, then the (A, B) intersection stores the same information as (B, A). On occasion, this may be desirable or undesirable. Naturally, a directed graph would store nonredundant information in each cell.
- Node Identities: The most logical choice for an underlying data structure would be a two-dimensional array. To leverage the constant-time lookups, we must provide an integer identifier for each node.
- Existence of Edges: Regardless of the points above, we must determine how to indicate that no edge exists. While many modern languages have some concept of a nullable type, you may not always want to use it. Particularly, nullable types often come with an implied increase in storage size (it may take 32 bits to store an integer, but storing an integer along with whether it exists is more information and consequently more bits). As a result, we might, by convention, choose a value to store in the matrix that indicates that no edge exists. Most weighted graphs in the natural world have strictly positive weights, so storing a –1 may serve as a useful indicator for a nonexistent edge. When working with an unweighted graph, simple 0s and 1s or true and false will suffice.

Adjacency Lists

If we perceive an adjacency matrix as a square table of edge information, an adjacency list is a jagged list of lists. The primary list has one entry for each node in the graph. Each of those entries then points to a list of adjacent nodes. The size of each secondary list depends on the degree of that node. Using the same graph as we used for adjacency matrices, we have the following adjacency list.

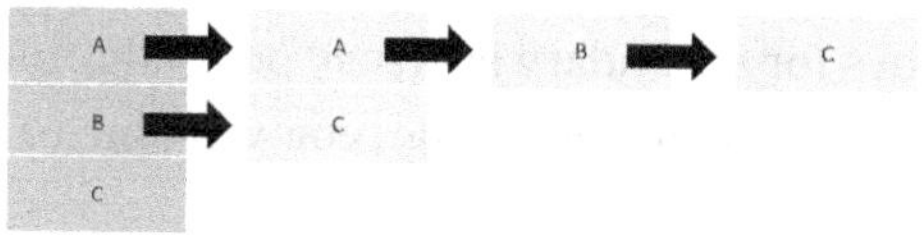

Figure 11.4

This is how adjacency lists are often portrayed visually, but we should note that the word "list" is in reference to the abstract data type list rather than a linked list. Also note that while the primary list clearly stores nodes, the secondary lists effectively store edges.

The list of lists nature of adjacency lists may be implemented in numerous ways. Considerations for concrete implementations include the following:

- Weighted or Unweighted: Weighted graphs require each entry in the secondary lists to store both the adjacent node's identity and that edge's weight. For this reason, we cannot store simple primitive types in each secondary entry. This implies that we

will likely need some kinds of composite types such as objects or structs. Unweighted graphs are easily stored using the identity of the node and may not require additional types.
- Directed or Undirected: As with adjacency matrices, undirected graphs tend to lead toward redundancy in data. If an undirected graph has an edge between A and B, then A's secondary list stores a reference to B, and B's secondary list stores a reference to A. Directed graphs have no such concern.
- Underlying Data Structures: As with adjacency matrices, it may be convenient to identify each node using an integer. This allows us to leverage constant-time lookups when looking for nodes in primary or secondary lists. Unlike adjacency matrices, using arrays for secondary lists may pose additional challenges if edges are frequently added or removed (due to the fixed size of arrays).

Algorithms

Much like our discussion of trees, graph algorithms could consume chapters of a textbook. Rather than a broad survey of problems and known algorithms, we will address only three specific problems.

Traversal

Two frequent questions with graphs are (1) how we can visit each node and (2) if we can find a path between two nodes. They arise whenever we want to broadcast on a network, find a route between two cities, or help a virtual actor through a maze. We rely on two related algorithms to accomplish this task: breadth first traversal and depth first traversal. If we wish to perform a search, we simply terminate the traversal once the target node has been located.

Both algorithms depend on knowing some start node. If we are attempting to traverse all nodes, a start node may be chosen arbitrarily. If we wish to find a path from one node to another, we obviously must choose our start node deliberately. Both algorithms work from the same basic principle: if we wish to visit every node originating from some start node, we first visit its neighbors, its neighbors' neighbors, and so on until all nodes have been visited. The primary distinction between both is the order in which we consider the next node. Breadth first spreads slowly, favoring nodes closest to the start node. Depth first reaches as deep as possible quickly. Below are examples of both traversals. Note that there is no unique breadth first or depth first traversal, but rather they are dependent on a precise implementation. The examples below represent possible traversals. There are other possibilities.

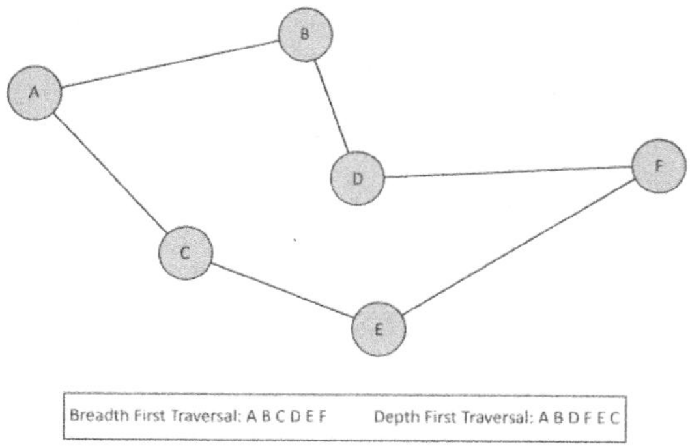

Figure 11.5

Note the difference in these traversals. Because breadth

first starting from A will consider *all* A's neighbors first, we will encounter C before we encounter F. Depth first will instead prioritize B's neighbors before completing all of A's neighbors. As a result, depth first reaches F before breadth first does.

In addition to the order in which we visit neighbors, we must also pay close attention to cycles. Recall that cycles are paths that start and end at the same node. In the depth first example above, what happens when we finally visit C only to find its neighbor is A? If we fail to recognize this as a cycle, we will again traverse A B D F E C and continue to do so indefinitely. We must avoid visiting already visited nodes. We will need to incorporate this into the algorithm as well.

Below is the pseudocode for a breadth first traversal. The function call **Visit** is simply a placeholder for some meaningful action you might take at each node (the simplest of which is simply printing the node identifier). It also assumes that node identifiers are integers.

```
function breadthFirst(startNode, graph, countOfNodes)
    set q to a new Queue
    set Visited to a Boolean array(length of countOfNodes)
    set all entries in Visited to false

    q.Enqueue(startNode)
    while (q is not empty)
        set n to q.Dequeue()
        if Visited[n] equals False
            Visit(n)
            set Visited[n] to True
            for each node m adjacent to n in graph
                q.Enqueue(m)
```

Consider the above pseudocode along with figure 11.5. Assume a mapping from the letters A–F to integers 1–6, respectively. We first enqueue A. The queue is not empty, so we dequeue A. We have not yet visited it, so we visit, mark as visited, then enqueue the neighbors (B and C). In the case of breadth first, these two nodes were enqueued before D, E, and F. As a result, B and C will be visited before D, E, and F.

Also note the utility of the visited array. We visit a node, mark it as visited, then enqueue its neighbors. Once we have visited a node and enqueued its neighbors, the conditional on line 9 will prevent us from doing the same again. This is our mechanism for avoiding cycles.

This pseudocode describes a breadth first traversal but only requires a nominal change to make it a depth first traversal. Recall that after we visited A, we visited B and C. This was due to the first-in-first-out nature of queues. Consider what happens if we swap our queue for a stack. We visit A and now push B and C. C was the last node pushed, so a pop returns it. We then push A and E, which will eventually be popped before B. As a result, we prioritize nodes deep in the graph before ever considering B. In fact, we eventually consider B due to its adjacency to D rather than its adjacency to A.

Three aspects of graph algorithms make runtime analysis difficult. Because of these challenges, runtime analysis in this chapter will be less precise than in others but still descriptive as to roughly how much effort is required to perform the task at hand.

- We are typically working with two variables: count of nodes and count of edges. In the case of breadth first traversal, we can see that we will only visit each node once. We also enqueue and dequeue once for each edge (plus an additional enqueue/dequeue for the start node). This gives us a runtime of O(N + E), where N is the number of nodes and E is the number of edges.
- Analysis can be confusing because the upper bound of E is $O(N^2)$. An explanation of why can be found in *Discrete Mathematics: An Open Introduction* (found in the references for this chapter). There the author explains how the number of edges in a complete graph relates to the number of nodes. The explanation closely resembles the justification for $O(N^2)$ runtime of selection and Insertion Sort. Because of these two aspects, we can correctly say O(N+E) or $O(N^2)$, the choice of which is typically dependent on the current context. If we

know that the number of edges is relatively low compared to the number of nodes, then use the sum of the two. If we know the graph to be highly connected, it is better to recognize the runtime as quadratic.

- Algorithms are typically presented conceptually without regard to precise implementations of graphs or auxiliary data structures. For example, if we are working in an object-oriented system, we may leverage adjacency lists, which limit our ability to perform constant-time indexing into the adjacent nodes. If we have no mapping from nodes to integers, we may have to perform Linear Searches to determine if nodes have been visited.

Single Source Shortest Path

While breadth first and depth first searches provide a path from a source to a destination, they do not guarantee an efficient path. Accomplishing such a task requires that the algorithm consider the weights of the edges that it traverses as well as the cumulative weights of edges already traversed. Numerous algorithms exist to find the shortest path from one node to another, but this section will focus on Edgar Dijkstra's algorithm.

Dijkstra's algorithm determines the shortest path between a source and destination node by maintaining a list of minimum distances required to reach each node already visited by the algorithm. It is an example of a greedy algorithm. This is a general strategy employed in algorithms, akin to the divide-and-conquer strategy employed in Binary Search or Merge Sort. Greedy algorithms make locally optimal choices that trend toward globally optimal solutions. In the case of Dijkstra's (and Prim's to follow), we only consider a single node at a time and use the information at that location in the graph to update the global state. If we carry this strategy out in clever ways, we can indeed determine the shortest path between two nodes without each step considering the entire graph.

In the following graph, consider finding the shortest path from A to D. Note that in the initial state, we acknowledge that the distance from the source node to itself is 0. This is analogous to enqueuing or pushing the source node in breadth first and depth first traversals. The primary control flow will again be a loop, which selects the next node to consider. Initializing some state again provides the loop with a logical place to begin. Also note that we maintain predecessors for each node whenever we update the distance to that node. This helps traverse the shortest path after the algorithm has been completed.

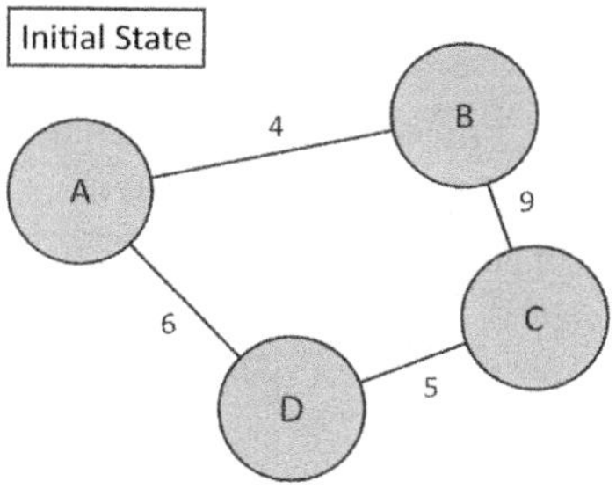

Node	A	B	C	D
Visited	F	F	F	F
Dist	0	Inf	Inf	Inf
Pre	Nil	Nil	Nil	Nil

Figure 11.6

Now that we have some predefined state, we will begin our iteration to update the distance and predecessor arrays with the best information we know so far. Given the nodes we have visited so far (namely, A), we know we can reach B with a weight of 4 and predecessor node A. We can also reach C with weight 6 and predecessor node A. Note here that we are not claiming that the edge AB is the shortest path from A to B or that AC is the shortest path from A to C. What we are claiming is that of the nodes visited so far, the minimum weight paths to B or C are 4 and 6, respectively. We then start the next iteration by carefully choosing the next node to visit. It should be one that has not yet been visited so that we do not create cycles. Additionally, regarding the shortest path, we must choose the next node based on which has the minimum distance from the starting node. We then repeat this process until each node is visited or we reach some desired destination node.

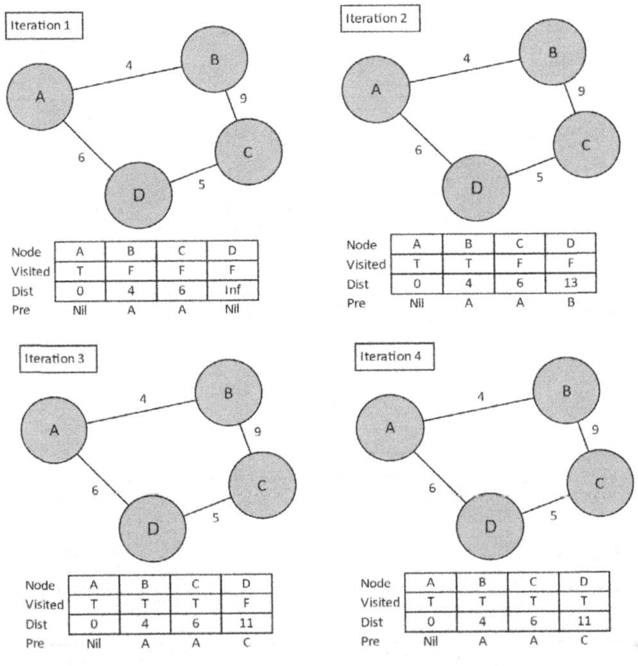

Figure 11.7

```
function Dijkstra(startNode, graph, countOfNodes)
    set startNode to visited and all others to not visited
    set startNode distance to 0 and all others to infinity
    set predecessors to nil

    while not all nodes marked as visited
        set n to node corresponding to minimum distance of unvisited nodes
        set n as visited
        for each unvisited node m adjacent to n in graph
            set e to edge from n to m
            if e.Weight + n.Dist < m.Dist
                set m.Dist to e.Weight + n.Dist
                set m.Pre to n
```

As with breadth first and depth first traversals, the precise runtime cannot be determined without specifying exactly how visited, distance, predecessors, and edges are structured. What we can determine with certainty is that the while-loop will iterate the same number of times as the number of nodes in the graph. This is evident because each iteration marks a node as visited, and the loop terminates when all nodes are visited. Assuming we are looking for all shortest paths (or our destination node is the last to be visited), we will perform the body of the inner loop once for each edge in the graph. This very closely approximates the runtime behavior of the traversal algorithms earlier and results in a likely worst-case runtime of $O(N^2)$. However, Dijkstra's algorithm is well researched, and known improvements can be made to this runtime by choosing clever data structures to represent different components.

Minimum Spanning Trees (MSTs)

A Minimum Spanning Tree (MST) is a subgraph (subset of edges) that satisfies the following conditions:

1. It must be a tree. In other words, there must be no cycles within the subgraph.
2. It must be spanning, which means that all the nodes in the original graph exist in the subgraph and are reachable using

only the edges in the subgraph.

3. It is possible to have more than one spanning tree for a given graph. Of those possible spanning trees, the MST is the one with the lowest cumulative edge weight.

As with other graph properties and algorithms, MSTs have numerous applications in the natural world. The canonical example is that of a network broadcast. Imagine computers as nodes and the network connections between them as edges. In computer networking, we often want to be able to broadcast a message to all nodes on the network (or, put simply, ensure that all nodes receive a particular message). The MST represents the lowest-cost means of transmitting such a message. Note that this is not the fastest means of transmission. That would indeed be a tree created by running a single-source shortest-path algorithm like Dijkstra's.

As with single-source shortest path, MSTs can be produced via numerous algorithms. The only one we address here is Prim's. We do so due to its similarity with Dijkstra's. Dijkstra's produced shortest paths by comparing a known distance against the sum of the cumulative distance to the predecessor plus the newly considered edge (see line 11 in the pseudocode). This comparison can be made because the distance for a node denotes the shortest path we have considered so far. Prim's algorithm changes the meaning of the distance value as well as the comparison on line 11. Rather than representing the cumulative distance as we did for shortest path, distance now represents the cost to add that node into the MST. The final change is simple: if we remove **n.Dist** from both lines 11 and 12, then we now find the MST rather than a shortest-path tree.

Exercises

1. Imagine you have a sparse graph with a large number of nodes but a relatively small number of edges. You are working in a system with strict constraints on how much memory you can consume. How does this impact your decision between an adjacency matrix and an adjacency list?
2. The breadth first example and pseudocode assume an unweighted and undirected graph. Does this pseudo code change if the graph is weighted? What if it is directed?
3. Consider an array of numbers. Devise a way of sorting these numbers using the graph algorithms from this chapter. The challenging part here is to determine how to model the array as a graph. **Hint:** What if the numbers are nodes, each node is connected to each other node, and the weight is the difference between the two nodes' values?

References

Levin, Ocsar. "Graph Theory." In *Discrete Mathematics: An Open Introduction*, 3rd ed., chap. 4. 2019.

http://discrete.openmathbooks.org/dmoi3/ch_graphtheory.html

12. Hard Problems

Learning Objectives

After reading this chapter you will...

- understand how computer scientists classify problems.
- be able to define some of the most common classes of problems in computer science.
- be able to explain the relationship between P and NP problem classes.
- understand some key properties of NP-complete and NP-hard problems.
- understand the role of approximate solutions and heuristics in battling hard problems.

Introduction

Everyone in life faces hard problems. Figuring out what to do with your life or career can be hard. You may find it hard to choose between two delicious menu items. These problems, while "hard" in their own way, are not the kinds of hard problems we will be exploring in this chapter. In this chapter, we will introduce some of the key ideas that support the theory of computation, the theoretical foundation of computer science. The discovery of these concepts is rather recent in the history of science and mathematics, but these concepts provide some fascinating insight into how humanity may attempt to solve the most difficult problems.

In the following sections, we will introduce the most discussed complexity classes in theoretical computer science. These are the complexity classes of P, NP, NP-complete, and NP-hard. These classes highlight many interesting and important results in computer science. We will explore what makes problems "easy" or "hard" in a theoretical sense. We will then explore some concepts for tackling these hard problems using approximations and heuristics. Finally, we will end the chapter with a discussion of an "impossible" problem, the halting problem, and what this means for computability.

The goal of this chapter is to simply introduce some of the important theoretical results in computer science and to highlight some ways in which this knowledge can be practical to a student of computer science. We will not introduce a lot of formal definitions or attempt to prove any results. This chapter is to serve as a jumping-off point for further study and, hopefully, an inspiring introduction to some of science's most profound discoveries about computing and problem-solving.

Easy vs. Hard

In some ways, what computer scientists view as an easy or hard problem is very simple to determine. Generally, if a problem can be solved in polynomial time—that is, $O(n^k)$ for some constant k—it is considered an easy problem. Another word that is used for this type of problem is "tractable." Problems that cannot be solved in polynomial time are said to be "intractable" or hard. These include problems whose algorithms scale exponentially by $O(2^n)$, factorially by $O(n!)$, or by any other function that grows faster than an $O(n^k)$ polynomial function. Remember though, we are thinking in a theoretical context. Supposing that k is the constant 273, then even with the small n of 2, 2^{273} is a number larger than the estimated number of atoms in the universe. In practice, though, few if any real problems have algorithms with such large degree polynomial scaling functions. By similar reasoning, some specific problem instances of our theoretically intractable problems can be exactly solved in a reasonable amount of time. In general, this is not the case though. Interesting problems in the real world remain challenging to solve exactly, but many of them can be approximated. These "pretty good" solutions can still be very useful. In the discussions below, we will mostly focus on time complexity, but a lot of theoretical study has gone into space complexity as well. Let's explore these ideas a bit more formally.

The P Complexity Class

In our discussion of hard problems, we need to first define some sets of problems and their properties. First, let's think about a problem that needs to be solved by a computer. A sorting problem, for example, provides an ordered list of numbers and asks that they be sorted. Solving this problem would provide the same numbers reordered such that they are all in increasing order. We know that there exist sorting algorithms that can solve this problem in $O(n^2)$ and even $O(n \log n)$ time. Problems such as these belong to the **P complexity class**. **P** represents the set of all problems for which there exists a polynomial time algorithm to solve them. This means that an algorithm exists for solving these problems with a time scaling function bounded by $O(n^k)$ for some constant k.

Strictly speaking, P is reserved only for **decision problems**, a problem with only a yes or no solution. This is not a serious limitation from our perspective. Many of the problems we have seen in this book can be easily reformulated as decision problems of equal difficulty. Suppose there is an algorithm, let's identify it as

A, that solves instances of a decision problem in P. If A can solve any instance of the problem in polynomial time, then we say that A **decides** that set of problem instances. For any input that is an instance of our problem, A will report 1. In this case, we say A **accepts** the input. If any input is not an instance of that decision problem, A will report 0. In this case, we say A **rejects** that input.

By framing our algorithms as decision problems, we can rely on some concepts from formal language theory. From this framework, we think about encoding our inputs as strings of 0 and 1 symbols. We should know numbers can be encoded in binary, but other types of data, such as images and symbol data, can also be so encoded. At some level, all their data are stored in binary on your phone or computer. We can use 0 and 1 as symbols to construct the strings of our binary language. In the formal language model, A acts as a language recognizer. If the input string is part of our specific language of problem instances, A will accept it as part of the language. If an input string is not part of the problem set of instances, A will reject it as we discussed in the previous paragraph. This is one of the formalizations that have been used to reason about problems in theoretical computer science. We will not explore formal languages any further here, but this model is equivalent to the practical problem-solving we have explored in this textbook. The language model also closely relates to the simplest theoretical model of computing, the Turing Machine.

The concept of **determinism** is another important idea to introduce in our discussion of the complexity class **P**. The **P** class is described as the class of **deterministic** polynomial time problems. This requires a bit of subtlety to describe accurately. For now, we will just say that the algorithms for solving problems in P function deterministically in a step-by-step fashion. This could be interpreted as meaning that the algorithms can only take one step at a time in their execution. This definition will make more sense as we discuss the next complexity class, **NP**, or the class of nondeterministic polynomial time problems.

The NP Complexity Class

We think of problems in P as being easy because "efficient" algorithms exist to solve them. By efficient, we mean having polynomial time complexity, $O(n^k)$. The **NP complexity class** introduces some problems that can be considered fairly hard. **NP** stands for **nondeterministic polynomial time** complexity. The NP class of problems introduces the idea of solution **verification**. If you were given the solution for an algorithm, could you verify that it was correct? Think about how you might verify that a list of numbers is sorted. How could you verify that 7! = 5040? I'm sure you can think of several ways to easily check these answers in a short amount of time. Again, we will focus on decision problems, but decision versions of all problems can be constructed such that we do not lose generality in this discussion. For a problem to be in **NP**, there must exist an algorithm A that verifies instances of the problem by checking a "**proof**" or "**certificate**." You may think of the certificate as a solution to the problem that must be verified in polynomial time.

NP leaves the question of whether a problem can be solved quickly and considers whether the solution could be verified quickly. The **nondeterministic** part refers to the idea of ignoring how quickly the problem could be solved. We mentioned that a deterministic algorithm could take only one step at a time. We could think of a nondeterministic algorithm as one that could take many steps "at the same time." One interpretation of this might be considering all options simultaneously. The main takeaway is that a correct solution must be verifiable in polynomial time for the problem to be a member of **NP**.

An Example of an NP Problem: Hamiltonian Cycle

At this point in our discussion, it may be helpful to examine a classic example of a problem in **NP**. A Hamiltonian cycle is a path in a graph that visits all nodes exactly once and returns to the path's start. Finding this kind of cycle can be useful. Consider a delivery truck that needs to make many stops. A helpful path might be one that leaves the warehouse, visits all the necessary stops (without repeating any), and returns to the warehouse. For the example graph below, we may wish to solve the decision problem of "Given the graph G = {V, E}, does a Hamiltonian cycle exist?"

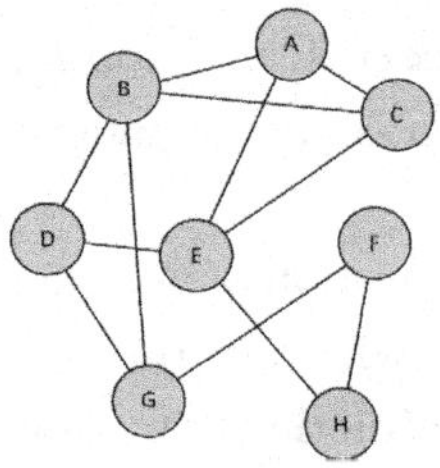

Figure 12.1

We will discuss the complexity of solving this problem soon, but for now, we will consider how to verify a solution to the problem. Suppose that we are given this problem and a potential solution. How would we verify the correctness of the solution? The "proof" or "certificate" of this problem could be the ordered list of vertices in the cycle. We could easily verify this solution by attempting to traverse the nodes (or vertices) in order along the graph. If we visit all the vertices and return to the starting vertex, the verification algorithm could report "yes." This would only require work proportional to the number of nodes, so *verifying* a solution to the Hamiltonian cycle problem would have a time complexity of O(n), where n is the number of nodes in the graph. This means that this problem could be easily verified, and by "easily," we mean it could be verified in polynomial time. For the above graph, a Hamiltonian cycle would be {E, A, C, B, D, G, F, H, E}. Note that we must return to the original position for the path to be a cycle. This is illustrated below:

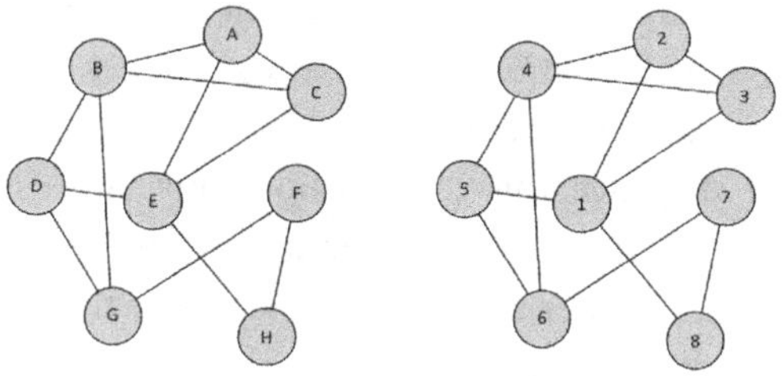

Figure 12.2

The fact that the Hamiltonian cycle problem can be easily verified may give the (false) impression that it is also easily solvable. This does not appear to be the case. One approach to solve it might be to enumerate all the possible cycles and verify each one. Each cycle would be some permutation of all the vertices. With n as the number of vertices in the graph, this means that there would be $O(n!)$ possible orderings to check! This naïve algorithm is even slower than exponential time $O(2^n)$. In fact, one of the best algorithms known to solve it has a runtime complexity of $O(2^n n^2)$—better than $O(n!)$ but still extremely slow for relatively small n.

Before we move on to the next section, let's consider the **P** complexity class in the context of **NP**. It should be clear that any problem in **P** must also be in **NP**. If a problem can be easily solved, it should also be easily verified. Consider for a moment the opposite situation where a problem is easy to solve but difficult to verify. Struggling to verify a solution to a problem might call into question how easily it was solved. The complexity class **P** represents all problems solved in polynomial time, and it is a subset of the **NP** class. Now whether it is a "proper subset" or not of **NP** is a classic unsolved problem in computer science theory. A proper subset means that it cannot be equivalent to the **NP** class itself. From this discussion, it may seem as though P and NP are not the same set, but many brilliant mathematicians and scientists have attempted to prove or disprove this fact without any success for decades. Whether P = NP or not remains unknown. In the next section, we will discuss this further and highlight just why the P = NP or P ≠ NP question is so interesting.

Polynomial Time Reductions

In this section, we will introduce the idea of a **reduction**. Informally, the term "reduction" refers to a method of casting one problem instance as an instance of another problem such that solving the new "reduced" problem also solves the original. As we explore the next two complexity classes of NP-hard and NP-complete, we use this powerful idea of reductions. Using an efficient reduction to transform one problem into another would serve as a key to solving a lot of different problems.

We will briefly consider a classic problem known as the **Circuit-Satisfiability Problem**. This is often abbreviated as **CIRCUIT-SAT**, but this could also represent the set of all circuit satisfiability problems (or, specifically, their instances). Suppose we want to determine if a circuit composed of logic gates has some assignment to its inputs that makes the overall circuit output 1. The circuits are composed of logic gates that take inputs that are either 0 or 1, standing for either low or high voltage. The typical diagram for these gates is given below:

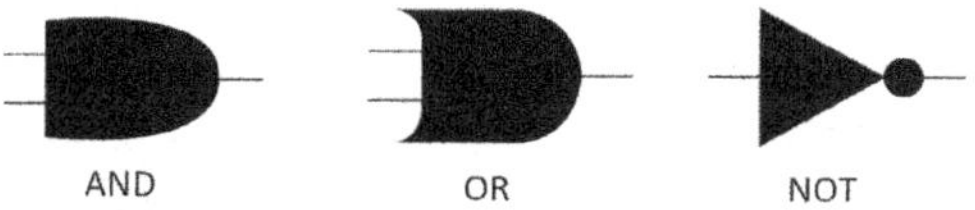

Figure 12.3

These gates correspond to their interpretation in mathematical logic. This means that the AND gate will output a 1 when both of its inputs are 1. We can compose these gates into larger circuits. The image below presents an example of a circuit that uses several of these gates and takes three inputs, marked X, Y, and Z:

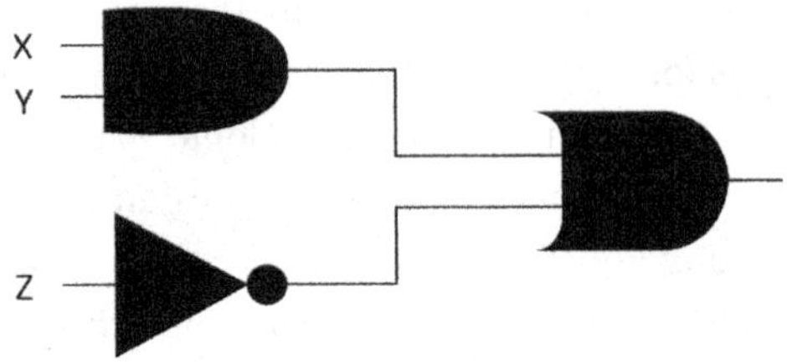

Figure 12.4

The decision problem for CIRCUIT-SAT would decide the question of "Given a representation of a circuit composed of logic gates, does an assignment of zeros and ones to the inputs exist that makes the overall circuit output 1?" Such an assignment of inputs is said to **satisfy** the circuit. One method of solving this problem would be to try all possible combinations of 0 and 1 assignments. Given n inputs, this would be attempting to try $O(2^n)$ possibilities. Given a potential solution, we could verify the assignment satisfies the circuit by simply simulating the propagation of input values through the sequence of logic gates. An algorithm for solving CIRCUIT-SAT problems would be very useful. Let's look at why.

Suppose we have another problem we wish to solve: Given a logical formula, can we provide an assignment to the logical Boolean variables that satisfies the formula? To satisfy the formula means to find an assignment of true or false values to the variables that makes the overall formula true. This is known as the **Boolean satisfiability** problem, and these problem instances are usually referred to as the set **SAT**. A logical formula can be composed of variables and Boolean functions on those variables. These are the functions AND, OR, and NOT. These are usually written as the symbols ^ (AND), ˅ (OR), and ¬ (NOT). Additionally, the formulas use parentheses to make sure there are no ambiguous connections. An example of a Boolean formula is given below:

$$(x \wedge y) \vee (\neg x \wedge z).$$

Formulas such as this can be used to model many problems in computer science. If we had an algorithm that could solve CIRCUIT-SAT problems, Boolean formula problems could be solved by first constructing a circuit that matched the formula and then passing that circuit representation to the algorithm that decides CIRCUIT-SAT. The figure below gives a circuit that corresponds to the Boolean formula given above:

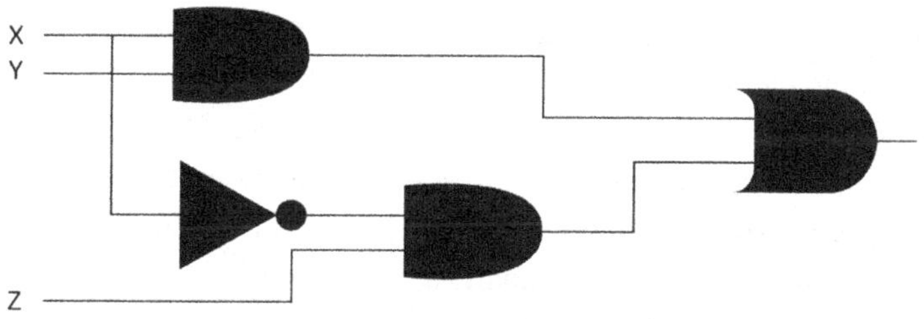

Figure 12.5

An assignment of 0 or 1 to the inputs of this circuit would correspond to an assignment of true or false to the Boolean variables of the formula. While not a formal proof, hopefully this illustration demonstrates how one instance of a problem can be cast into another and a solution to one can be used to solve the other. A key point is that this conversion must also be efficient. For this strategy to be effective, the *reduction from one problem* (SAT) *to another problem* (CIRCUIT-SAT) *must also be efficient*. If just doing the reduction was intractable and difficult, then we would not make any progress. We will only be interested in reductions that can be done in polynomial time. For this problem, we could create a procedure that would parse a string representation of the formula and generate a parse tree. From this tree, we could use each branching node to represent a logic gate, and from this, we could construct a representation of the circuit. Generating the parse tree might require $O(n^3)$ operations (this is an upper bound on some parsing algorithms), and converting the tree could be done using a tree traversal costing $O(n)$. This means that for this case, we could efficiently "reduce" the SAT problem into an instance of CIRCUIT-SAT.

We will introduce the notation for reducibility here, as it will be helpful in the following discussions. Remember that we can also talk about the representations of problems as being strings in a language. We might say that SAT, or all the problems in SAT, represents a language L_1. The problems in CIRCUIT-SAT represent the language L_2. Now to capture the above discussion in this notation, we would write $L_1 \leq_P L_2$, using a less than or equal to symbol with a P subscript. The meaning of $L_1 \leq_P L_2$ is that L_1 is **polynomial-time reducible** into an instance of L_2. The less than or equal to symbol is used to mean that problems in L_2 are at least as hard as problems in L_1. The P subscript is there to remind us that the reduction must be doable in polynomial time for this to be a useful reduction.

Let's provide one more example of a reduction. Another interesting and well-studied problem in computer science is the **Traveling Salesman Problem** or **TSP**. This problem tries to solve the practical task of minimizing the amount of travel between the different cities for a salesperson before they return home. Another way to cast the problem might be to ask, "What is the route that minimizes energy usage for a delivery truck such that it makes all its stops and returns to the warehouse?" You may already be thinking back to our discussion of Hamiltonian cycles. The TSP is looking for a minimum-cost **tour**, which is precisely a Hamiltonian cycle. To consider the decision version of the TSP, we would take a graph with edge weights representing the costs of traveling from one destination to another and a cost threshold k. The decision problem then asks, "Given the weighted graph G and the threshold k, does there exist a minimum cost tour with a cost at most k?" So an instance of the Hamiltonian cycle problem could be reduced to an instance of the TSP. Taking an instance of the Hamiltonian cycle problem, we could construct a new graph with all edge weights set to 0. This could be done easily in polynomial time by modifying the representation of the graphic. This new weighted graph could be passed to an algorithm from solving TSP with k set to 0. Let's let the set of all instances of Hamiltonian cycle problems be HAM-CYCLE. This means that we have HAM-CYCLE $\leq_P$ TSP, and any algorithm that solves instances of TSP can solve instances of HAM-CYCLE.

The NP-Hard and NP-Complete Complexity Classes

Reductions serve as a key to solving problems by taking them from one type of problem and transforming them into another. We explored two examples of reductions in the previous section. The SAT problems are reducible to the CIRCUIT-SAT problems. The HAM-CYCLE problems are reducible to the TSP problems. Other clever results have demonstrated that three-coloring a graph is reducible to the SAT problems. Interestingly, there are algorithms that can solve any problem in **NP** by reducing them from other problem types into an instance of a specific **NP** problem. These problems represent the **NP-hard** complexity class. More formally, an **NP-hard** problem is a problem (language) L, such that for any problem L′ in **NP**, L′ $\leq_P$ L. In other words, any algorithm for solving an **NP-hard** problem could solve ***any*** problem in **NP**. All four of our problems—CIRCUIT-SAT, SAT, HAM-CYCLE, and TSP—are **NP-hard**. The **Cook–Levin theorem** proved an interesting result showing that SAT is both in **NP-hard** (can be used to solve any **NP** problem) and in **NP** (easily verifiable). The class of problems with these characteristics is known as the **NP-complete** problems.

Now we revisit the P = NP or P ≠ NP question. Why is this a big deal? Suppose a problem set (and algorithm) could be found that was in **NP-complete** and in **P**. This would mean we have an **NP-hard** problem that can be easily solved. This result would mean that any **NP** problem could be easily solved in $O(n^k)$ time. We would simply reduce any **NP** problem into an instance of our special problem and solve it in polynomial time. This scenario would be the incredible result of a **P = NP** reality. The question is still up for debate, and no one has been able to prove this fact or, more importantly, find the algorithm. A world in which all difficult problems could be easily solved would certainly be interesting. For now, it is unknown whether P = NP or P ≠ NP. Many believe that P ≠ NP is the more likely scenario, but it has never been proven.

Approximation Algorithms and Heuristics

We should discuss the practical matter of how to solve difficult problems. We have given a somewhat formal description of NP-Hard and NP-Complete complexity classes, but let's reconsider these problems in practical terms. Suppose we need to solve a SAT problem with 60 variables, and we **brute-force** search by trying every combination of Boolean assignments and evaluating them. The brute-force search requires $O(2^n)$ operations. So with 60 variables, the number of combinations is on the order of 2^{60}. We call the set of all possible solutions the **search space**. If we assume a computer could check 2 billion of these possible assignment solutions per second (which is reasonable), we could expect the calculation to be completed in about 18 years. The worst-case exponential time complexity for exploring the search space means that solving these problems quickly is impossible even for relatively small n (< 100).

We want solutions very quickly and cannot wait 18 years to figure out our best delivery route for this morning's deliveries. Delivery companies want to be efficient to conserve energy. Factories want to maximize output and keep their machines running. Sometimes a great **approximate solution** to an **NP-complete** problem can be found quickly. An approximate solution is not totally correct, but it may satisfy many of the problem's requirements. Suppose that we found a SAT assignment that could satisfy most of our Boolean formula's expressions in the previous example; then this might still be very useful. Many real-world problems can be modeled by **NP-complete** problems, so finding good approximations for them is important work.

Many strategies exist for finding good approximations. The search for a good approximation can be framed as an optimization problem. We want to optimize a current solution's value toward the optimal value of a fully correct solution. One approach might be to randomly try many different solutions and calculate the value. Each time you find a solution with a better value, you save it as the current best. You let the algorithm run for a fixed amount of time. When the time is up, return the best solution that was found. In general, the search for a good approximation makes use of **heuristics**. Heuristics are strategies or policies that help direct a search algorithm toward better approximations. The hope is that the heuristic will help guide the search toward an optimal solution. Unfortunately, this is not a guarantee. Algorithms usually act on local information, so any heuristic might be guiding the search toward a **local optimum** while the global optimum is in the other direction. Developing heuristics for **NP-complete** problems is an active field of research. We will look at one heuristic, the **greedy algorithm**, and see how it might be applied to an NP-Hard problem.

The greedy algorithm uses the heuristic to always make the choice that maximizes the current value. To explore this heuristic, we will introduce another **NP-hard** problem. The **bin packing problem** seeks to optimally pack objects of different sizes into a fixed-size bin. Each item has a cost associated with it, and the bin has a capacity threshold where no items may be added that would push the total cost over the threshold. You can think of this as the bin getting full of stuff, and nothing else can be put in it. The example below gives an illustration of the bin packing problem:

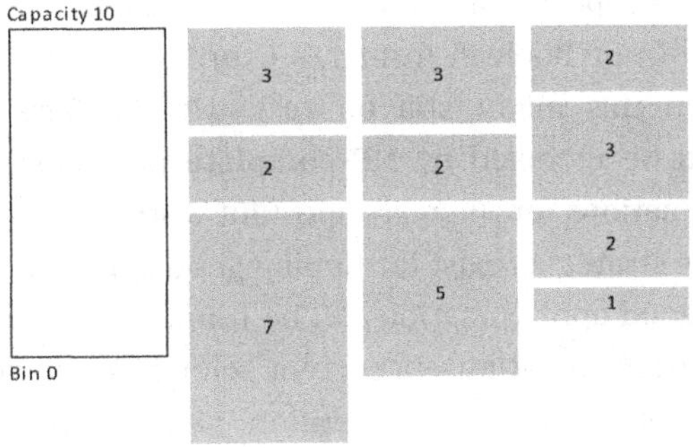

Figure 12.6

Given the boxes and their sizes, is there a way to pack all the boxes in the minimum number of bins? It might seem simple, but to solve this problem optimally, in general, might require a lot of time. One approach to finding the optimal number of bins would be to try all orderings of the items. Attempt to create bins by taking the items in the ordering and opening a new bin when the first is full. By trying all possible orderings of the items, the optimal bin number would be found, but this would take O(n!) time.

Using the **greedy heuristic** may help speed up our search even if the result may be suboptimal. A **greedy algorithm** tries to maximize or minimize the current value associated with a solution. For bin packing, a greedy strategy would be to always put the current item in the bin that minimizes the bin's extra capacity. In other words, put the item in the bin where it fits the tightest. This is known as the **Best Fit** algorithm. An example of a Best Fit solution is presented below for the ordering {3, 3, 2, 3, 1, 2, 2, 5, 7, 2}. This assumes that the items arrive in a fixed order, and they cannot be reordered. We do get to choose which bin to place them in though. This is sometimes known as the "online" version of the bin packing problem.

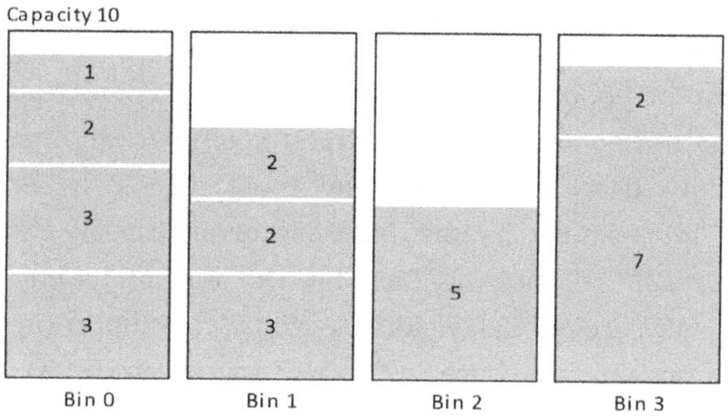

Figure 12.7

At each step, the algorithm tries to create the most tightly packed bin possible. A clever algorithm for Best Fit achieves an O(n log n) time complexity by querying bins by their remaining capacity in a balanced binary search tree. This algorithm is extremely fast compared to the brute-force method, but it is not optimal. Below is an optimal solution:

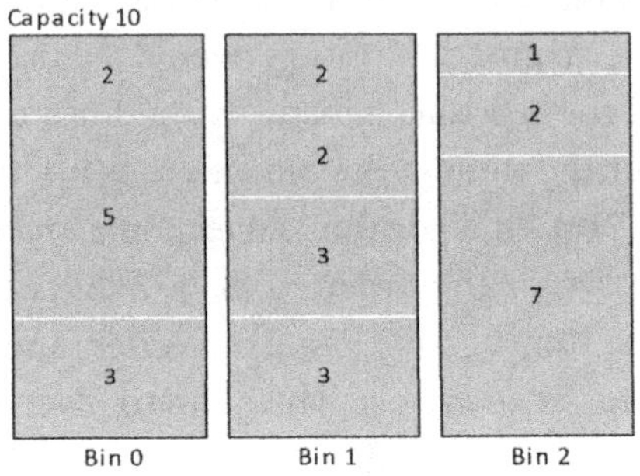

Figure 12.8

Depending on whether the items can be reordered or not, we may have the opportunity to first sort the items before applying Best Fit. Another good greedy algorithm for bin packing first sorts the items into descending order and then applies the Best Fit algorithm. This is known as **Best Fit Decreasing**. The figure below shows the result of applying Best Fit Decreasing to our block problem. This strategy does yield an optimal solution in this case. This algorithm would also have an O(n log n) time complexity. These algorithms show the value of using a heuristic to discover a good approximate solution to a very difficult problem in a reasonable amount of time.

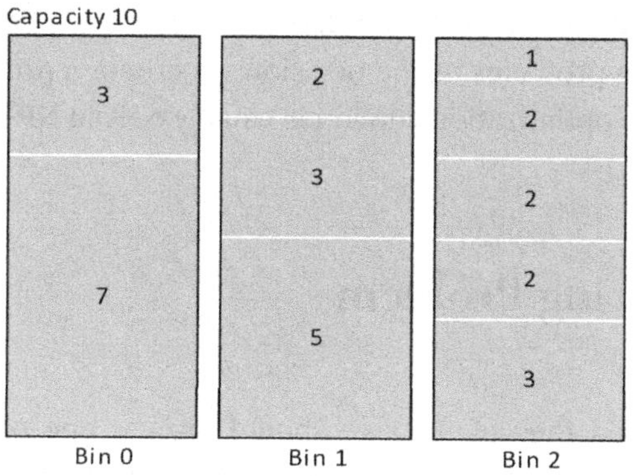

Figure 12.9

Bin packing provides insight into another feature of **NP-complete** and **NP-hard** problems. The decision version of the bin packing problem asks, "Given the n items and their sizes, can all items be packed into k or fewer bins?" This decision problem turns out to be NP-Complete. Given a potential solution and the number of bins, we can easily verify the number of bins used and the excess capacity in O(n) time. This fact confirms that the problem is in **NP**. Even with a target number of bins given, we would have to try overwhelmingly many configurations to ultimately determine if all the items would fit into the k bins. Now we may also be interested in determining the optimal number of bins. This decision problem might be asked as "Given the n items and their sizes, does the minimum packing require at most k bins?" Consider how we might verify that the optimal configuration was found. This means that we were given a solution and told it is optimal. We would now need to verify it. We could easily verify if the solution fits into the given number of bins. On the other hand, verifying that the number of bins for this solution is optimal would require considering all the possible solutions and checking that no other solution exists with a smaller number of bins. This means that the optimization version of this problem is not in **NP**. Therefore, the optimization problem is only **NP-hard** and not **NP-complete**. This pattern is common with **NP-complete** problems. If the decision version of a problem is **NP-complete**, its optimization version is usually only in **NP-hard**.

The Halting Problem

Before we end the chapter, we should discuss one of the classic problems in computer science, the **halting problem**. The halting problem illustrates the existence of "unsolvable" problems. Alan Turing proved the existence of a particular **undecidable** problem. The halting problem can be defined as asking the question "Given a representation of a computer program and the program input, will the program halt for that given input or run forever?" Turing's argument proposed the existence of a program (an algorithm running on a machine) that could detect if another program would halt given a specific input. Let's just informally say we have a function like **checkIfHalts(program, input)**. If the input program would halt, meaning complete successfully, on the given input, then **checkIfHalts** would report yes. If the program would run forever given the input, **checkIfHalts** would report no. This would be an algorithm that decides the halting problem. Running this hypothetical algorithm on a machine would allow a scheme for deciding if a program would halt. The program **checkIfHalts** would simulate the program P with the given input and decide if P halts on the input. This machine is presented in a diagram below:

checkIfHalts (program, input)

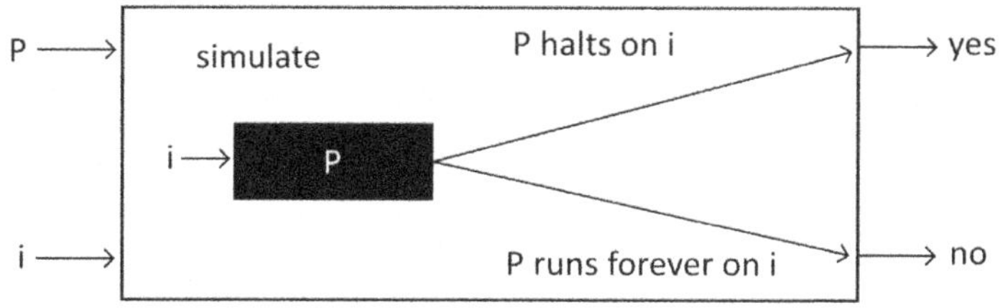

Figure 12.10

The interesting part of the argument suggests that our program runs with its own representation presented as the input. Let's construct another machine using **checkIfHalts** that will run forever if the program halts given an input but will halt if the program runs forever (as verified by **checkIfHalts**). Below is a diagram of this machine. We will call it **loopIfHalts**.

loopIfHalts (program, input)

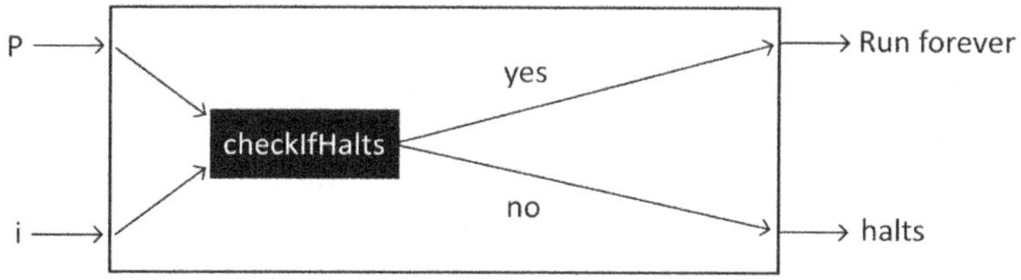

Figure 12.11

Now we construct one final machine as follows. This machine will take as input the representation of a program and try to determine if that program would halt when given a representation of itself as input. This is done by copying the program and using the copy as input. This machine is presented below. We will just call this **M(program)**.

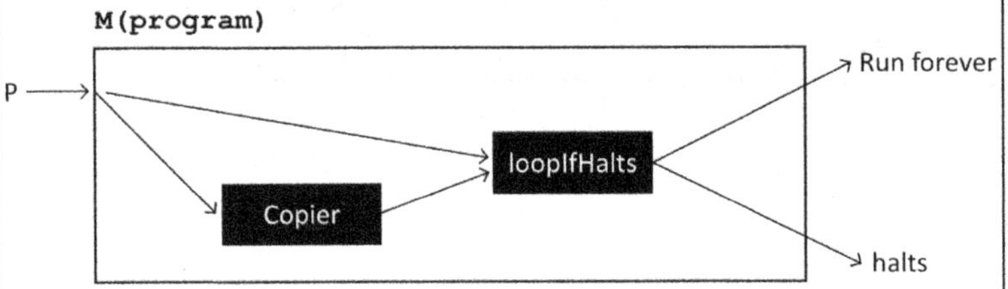

Figure 12.12

Now suppose that we run M with a representation of M as the input. We can think of this as calling M(M). If the program M should halt given M as the input, then M(M) should run forever. However, this is exactly what we did. We passed M into M, and if it runs forever, then M(M) should halt. This leads to a contradiction. We have a paradox where M should both run forever and halt. Since we arrived at a contradiction and all these algorithms (**loopIfHalts** and M) are derived from our hypothetical **checkIfHalts** program, these facts indicate that such a program cannot exist. This means that the halting problem is **undecidable**. This proof was discovered by Alan Turing and published in 1936. It provided some of the first evidence of problems that were literally unsolvable. Now that's a hard problem!

Summary

In this chapter we have explored many incredible results in computer science theory. We have explored what it means for a problem to be hard in a practical sense and in a theoretical sense. We also discussed the existence of impossible problems. These problems cannot be solved by any computer no matter how powerful or how much time they are given. Scientists are still working to understand whether P = NP or not. If this result turns out to be true, there may exist an efficient algorithm for solving many of our most difficult problems exactly. For now, though, no such algorithm is known. Computer scientists and humans in general never give up in the face of hard problems. We also explored the use of heuristics to help find suitable solutions when an exact solution might not be practical to find. These results in computer science theory will help you understand what makes problems hard and what to do about them.

Exercises

1. Do some research on NP-complete problems. Find an NP-complete problem that was not discussed in this chapter. What is the current best time complexity for the problem?
2. For your problem in exercise 1, how efficient in terms of runtime complexity are the current best approximation algorithms for the problem? What

heuristics are used in the approximate solution?

3. In your language of choice, implement the Best Fit algorithm for bin packing. Feel free to use a Linear Search rather than a balanced search tree. Use an interactive loop to allow the user to enter different sizes for each of the items and apply the greedy algorithm. Compare your implementation results to examples from this chapter.

4. Try the following thought exercise. Consider the possibility that an algorithm is discovered that solves NP-complete problems in polynomial time. Write a paragraph describing how our society might change with the advent of this algorithm. Be sure to address some specific algorithms that could be made efficient and how solving them quickly might impact society.

References

Bellman, Richard. "Dynamic Programming Treatment of the Travelling Salesman Problem." *Journal of the* ACM (JACM) 9, no. 1 (1962): 61–63.

Cormen, Thomas H., Charles E. Leiserson, Ronald L. Rivest, and Clifford Stein. *Introduction to Algorithms*, 2nd ed. Cambridge, MA: The MIT Press, 2001.

Held, Michael, and Richard M. Karp. "A Dynamic Programming Approach to Sequencing Problems." *Journal of the Society for Industrial and Applied Mathematics* 10, no. 1 (1962): 196–210.

Tovey, Craig A. "Tutorial on Computational Complexity." *Interfaces* 32, no. 3 (2002): 30–61.

Turing, Alan Mathison. "On Computable Numbers, with an Application to the Entscheidungsproblem." *J. of Math.* 58, no. 5(1936): 345–363.

Contributors

Authors

Paul W. Bible
DEPAUW UNIVERSITY

Paul W. Bible is currently a faculty member in the Department of Computer Science at DePauw University in Greencastle Indiana. He holds a Ph.D. in Computer Science and has conducted research in bioinformatics and computational biology both internationally and at the National Institutes of Health. Dr. Bible believes in the power of equity in education to drive social change. He hopes that this book will help more students succeed on their path to becoming computing professionals.

https://orcid.org/0000-0001-9969-4492

Lucas Moser
MARIAN UNIVERSITY

Lucas Moser is an independent consultant and faculty member at Marian University's Department of Mathematical and Computational Science. There he passionately shares his assertion that a rich education plays a major role in the development of problem-solving skills. His experiences in software engineering, management, and teaching bring a unique perspective to both project teams and students.

https://orcid.org/0009-0006-9452-1246

Illustrator

Mia M. Scarlato

Reviewers

Aaron Boudreaux
UNIVERSITY OF LOUISIANA AT LAFAYETTE

Joshua Kiers
MARIAN UNIVERSITY

Made in the USA
Monee, IL
04 November 2026

67380162R00193